Daniel Gurski

Customer Experiences affect Customer Loyalty

An Empirical Investigation of the Starbucks Experience using Structural Equation Modeling

Anchor Compact

Gurski, Daniel: Customer Experiences affect Customer Loyalty: An Empirical
Investigation of the Starbucks Experience using Structural Equation Modeling.
Hamburg, Anchor Academic Publishing 2013
Original title of the thesis: How Excellent Experiences affect Customer Loyalty

Buch-ISBN: 978-3-95489-118-4
PDF-eBook-ISBN: 978-3-95489-618-9
Druck/Herstellung: Anchor Academic Publishing, Hamburg, 2013
Additionally: Niederlande, Maastricht University, School of Business and Economics

Bibliografische Information der Deutschen Nationalbibliothek:
Die Deutsche Nationalbibliothek verzeichnet diese Publikation in der Deutschen
Nationalbibliografie; detaillierte bibliografische Daten sind im Internet über
http://dnb.d-nb.de abrufbar

Bibliographical Information of the German National Library:
The German National Library lists this publication in the German National Bibliography.
Detailed bibliographic data can be found at: http://dnb.d-nb.de

All rights reserved. This publication may not be reproduced, stored in a retrieval system
or transmitted, in any form or by any means, electronic, mechanical, photocopying,
recording or otherwise, without the prior permission of the publishers.

Das Werk einschließlich aller seiner Teile ist urheberrechtlich geschützt. Jede Verwertung
außerhalb der Grenzen des Urheberrechtsgesetzes ist ohne Zustimmung des Verlages
unzulässig und strafbar. Dies gilt insbesondere für Vervielfältigungen, Übersetzungen,
Mikroverfilmungen und die Einspeicherung und Bearbeitung in elektronischen Systemen.

Die Wiedergabe von Gebrauchsnamen, Handelsnamen, Warenbezeichnungen usw. in
diesem Werk berechtigt auch ohne besondere Kennzeichnung nicht zu der Annahme,
dass solche Namen im Sinne der Warenzeichen- und Markenschutz-Gesetzgebung als frei
zu betrachten wären und daher von jedermann benutzt werden dürften.

Die Informationen in diesem Werk wurden mit Sorgfalt erarbeitet. Dennoch können
Fehler nicht vollständig ausgeschlossen werden und die Diplomica Verlag GmbH, die
Autoren oder Übersetzer übernehmen keine juristische Verantwortung oder irgendeine
Haftung für evtl. verbliebene fehlerhafte Angaben und deren Folgen.

Alle Rechte vorbehalten

© Anchor Academic Publishing, ein Imprint der Diplomica® Verlag GmbH
http://www.diplom.de, Hamburg 2013
Printed in Germany

To Mom and Dad.
Thanks for supporting me in everything I do.
You are wonderful parents.

Special thanks also to Dr. Néomie Raassens and
Prof. Dr. Gaby Odekerken-Schröder for
the helpful support and critical input.

Table of Contents

List of Figures VII
List of Tables VIII
List of Abbreviations IX
1. Introduction 1
2. Literature Review 3
   2.1 The Evolution from Products to Services to Experiences 4
   2.2 The Initial Conceptual Model 6
      2.2.1 Customer Experience Quality 7
      2.2.2 Perceived Value 11
      2.2.3 Customer Loyalty 11
      2.2.4 Perceived Wealth 12
3. Methodology & Research Design 15
   3.1 Assigning Scales to the Individual Constructs 17
   3.2 Pre-Testing the Scales 18
      3.2.1 Data Collection 18
      3.2.2 Scale Purification Process 18
         3.2.2.1 CXQ scale 19
         3.2.2.2 Perceived Wealth scale 21
         3.2.2.3 Perceived Value scale 22
         3.2.2.4 Customer Loyalty scale 22
   3.3 Adjustments and Refinements 23
   3.4 Testing the Measurement Model 24
      3.4.1 Data Collection 24
      3.4.2 Measurement Model 26
      3.4.3 Assessing Model Fit of the Measurement Model 27
      3.4.4 Assessing Validity of the Measurement Model 30
4. Data Analysis 31
   4.1 Comparison of Competing Models 33
      4.1.1 Interpretation of Structural Model #1 suggesting Full Mediation 34
      4.1.2 Interpretation of Structural Model #2 suggesting Partial Mediation 36
      4.1.3 Interpretation of Structural Model #3 suggesting No Mediation 38
   4.2 Selection of the Best Fitting Model 39
5. Discussion 40
6. Conclusion 41
   6.1 Theoretical Implications 41
   6.2 Managerial Implications 42
   6.3 Limitations & Future Research 44
Reference List 46
Appendix 52

List of Figures

Figure 1. Price of coffee offering from commodity to experience.. 2
Figure 2. Progression of Economic Value.. 5
Figure 3. Initial Conceptual Model 6
Figure 4. Refined Conceptual Model after Pre-Test 23
Figure 5. Measurement Model #1.0 in AMOS.. 28
Figure 6. Final Measurement Model #1.2 in AMOS 32
Figure 7. Structural Model #1 35
Figure 8. Structural Model #2 37
Figure 9. Structural Model #3 38

List of Tables

Table 1. Existing Conceptualizations of Customer Experience Quality 9
Table 2. Dimensions of Customer Experience Quality 10
Table 3. Results of Factor and Reliability Analysis for the CXQ scale 20
Table 4. Results of Factor and Reliability Analysis for the Product Quality scale 21
Table 5. Results of Factor and Reliability Analysis for the Perceived Wealth scale 21
Table 6. Results of Factor and Reliability Analysis for the Perceived Value scale 22
Table 7. Results of Factor and Reliability Analysis for the Loyalty Intention scale............... 22
Table 8. Model Fit of Measurement Models 28
Table 9. Standardized Total Effects in Measurement Model #1.1 30
Table 10. Model Fit of Structural Models (Full Dataset) 32
Table 11. Multi-group effects of Perceived Wealth in Structural Model #1 (e5 fixed) 35
Table 12. Multi-group effects of Perceived Wealth in Structural Model #2 (e5 fixed) 37
Table 13. Multi-group effects of Perceived Wealth in Structural Model #3 (e5 fixed) 39

List of Abbreviations

AGFI	Adjusted Goodness-of-Fit Index
AQ	Atmosphere Quality
CFA	Confirmatory Factor Analysis
CFI	Comparative Fit Index
df	Degrees of Freedom
EFA	Exploratory Factor Analysis
FQ	Flow Quality
HIT	Human Intelligence Task
LQ	Learning Quality
MTurk	Amazon Mechanical Turk network
n	Sample Size
PCFI	Parsimony Comparative Fit Index
PQ	Product Quality
PRATIO	Parsimony Ratio
RMSEA	Root Mean Square Error of Estimation
SEM	Structural Equation Modeling
SQ	Service Quality
WOM	Word-Of-Mouth

1. Introduction

The world has changed. Customers' increasing expectations towards companies make competition continuously harder. What it needs are new business strategies that attract customers and, even more important, business strategies that make customers loyal. This is not a new phenomenon but rather a continuous evolution in business that especially gained momentum in the last years. Over time a development took place from products to services to a post-product, post-service phenomenon which is still evolving (Maklan and Klaus, 2011).

Marketers came to the point where they realized that striving for mere customer satisfaction might not be the panacea to create customer loyalty as it was expected to be. More than 60% of customers who switch to another brand identify themselves as satisfied (Jones, 1996; Reichheld, 1993). Regardless of these academic insights, most companies still rely on it. A report by Euromonitor International based on observations in the American market acknowledges the necessity for marketers to reset strategies. "Provide not only tangible products, but also unforgettable experiences!" (Euromonitor, 2008). In recent years, customer experiences as the ultimate competitive element increasingly gained attention among practicioners and theorists (e.g., Maklan and Klaus, 2011; Verhoef et al., 2008; Gentile et al., 2007; LaSalle and Britton, 2003; Carù and Cova, 2003; Pine and Gilmore, 1998).

It is commonly acknowledged that consumption activities almost always contain experiential aspects (Holbrook and Hirschman, 1982) and it should be the company's vital mission to make them extraordinary and compelling in order to differentiate from competition and gain a competitive advantage (Pine and Gilmore, 1998). In 2013 we can already find a customer experience focused mindset in some companies' mission statements (e.g., Dell: "Dell's mission is to be the most successful computer company in the world at delivering the best customer experience in markets we serve.") or core values (e.g., McDonald's: "We place the customer experience at the core of all we do.").

An often cited example of experience staging and its financial benefits for the company is the American Coffee Company Starbucks. During the transformation from a commodity (=harvested coffee beans) into a good (=roasted, grinded, packaged coffee beans) into a service (=simply served cup of brewed coffee) and finally into an experience (=extraordinary way of ordering, creation, and consumption of a cup of coffee) the price charged increases exponentially (Pine and Gilmore, 2011). Figure 1 shows the different price ranges that are paid on average by customers in the distinct "evolutionary stages" for a cup of coffee in the United States.

Figure 1. Price of coffee offering from commodity to experience (Pine and Gilmore, 2011)

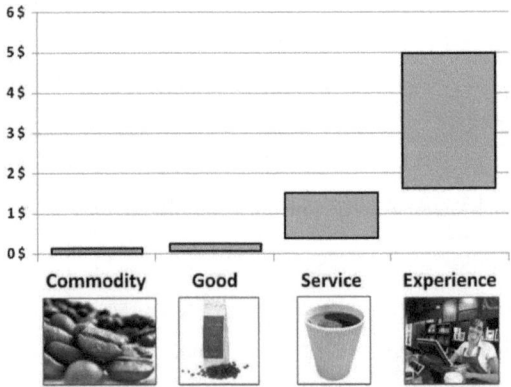

We see that it is crucial for managers to extend their mindset beyond product quality and service quality. They need to manage the experience their company offers. But this is easier said than done. The American statistician William Edwards Deming once said "You cannot manage what you cannot measure" and that explains nicely the current shortcoming of customer experience literature. Although "customer experience" has become an omnipresent buzzword in marketing, it still has severe theoretical shortcomings in terms of academic research. In the existing literature, the importance of a successful management of the customer experience is indeed emphasized, but researchers often fail to measure customer experiences holisticly and if they do, like Lax (2012) also recognizes, they usually fail to link customer experience quality to the larger context of business outcomes and customer loyalty. Hence, it is hard to manage customer experiences on the basis of existing models. What is needed as a proper foundation is a conceptual framework with a set of meaningful measurement scales that links the quality of a customer experience to customer behavioral outcomes, especially customer loyalty.

Therefore, the study at hand aims to overcome the aforementioned pitfalls and presents a model that 1) holisticly measures customer experience quality (with the newly developed CXQ scale) and 2) links the different dimensions of customer experience quality to customer loyalty intentions in the form of word-of-mouth and the customer's willingness to pay more. 3) In order to gain valuable insights for customer segmentation, perceived wealth is introduced as a moderating variable in the model. The study hereby primarily investigates the effect of customer experience quality on the different facets of customer loyalty.

In chapter two, the conceptual model will be presented and acts as a framework for the discussion on existing literature regarding customer experiences, including its implications and gaps. On this basis the hypotheses will be developed.

Chapter three describes the research design and methodology used to empirically test the conceptual model and hypotheses. Structural Equation Modeling (SEM) will be applied in this study to validate the model and test the hypotheses. Based on common sense and the literature discussed in chapter three, the proposed scales to measure the distinct constructs will be presented. Via exploratory factor analysis the scales will be purified and refined to be afterwards transferred into a measurement model which is tested via confirmatory factor analysis. As soon as the fit, reliability and validity of the measurement model is assessed, the relationships as stated by the hypotheses will be integrated into a set of competing structural models and tested in chapter four.

In chapter five, explanations for the observed findings will be discussed critically. During the conclusion in chapter six, the author outlines the theoretical and managerial implications in detail, discusses limitations of the study, and suggests topics for future research.

In advance, the findings of this study are valuable for both theory and practice. It will bring companies one major step forward towards the successful management of customer experiences and allows companies to stage them efficiently and effectively in order to gain a significant competitive advantage. Other researchers can use the CXQ measurement scale for future research and introduce new moderators in order to advise companies in terms of customer segmentation. Additional behavioral outcome or business outcome variables which are integrated in the model might lead to valuable insights as well.

2. Literature Review

Customer experience is a popular, but equivocal marketing buzzword. There are as many different academic conceptualizations as there are scholars and measuring it is a challenge due to its latent nature. Although the literature regarding customer experiences already evolved three decades ago from articles such as Hirschmann and Holbrook (1982), dealing with experiential aspects of consumption, literature still fails to provide a precise terminology and standardized, generalizable approaches (Gentile et al., 2007). To provide the reader with a profound overview of the topic without losing ourselves in the width of diverging definitions, the following literature review will be structured as follows: First, the author will describe the ongoing process in today's business to increasingly focus on experiences as a progression from products and services. Second, the author presents a conceptual model that combines insights of different studies from the fields of marketing, psychology as well as financial economics. The conceptual model serves as a framework to discuss the different variables, logically relates them to each other based on previous research and visualizes the hypotheses which are consequtively tested in chapter three, four and five.

2.1 The Evolution from Products to Services to Experiences

Over the past three decades, marketing theory underwent several large-scale paradigm shifts. The described paradigm shifts in theoretical literature reflect real-world changes in the competitive landscape that companies face. Maklan and Klaus (2011) conclude that a development from products to services to a post-product, post-service phenomenon took place which is still evolving. It is important to emphasize that the requirements for companies were not substituted over the different phases but rather extended into new dimensions, making competition more and more complex.

In the 1990s the first paradigm shift occurred when marketers took the relational aspects between customer and company into account (Grönroos, 1997; Christopher, 1996). Before, the classic product marketing was largely focused on sales and the creation of fast moving consumer good brands (Merz and Vargo, 2009; Copeland, 1923). Now, instead of bringing products "to market" and considering consumers as targets, firms started to co-create value collaboratively and "marketed with" their customers over an extended time frame. Instead of "delivering value" like in previous eras, the firm's role was now seen as "proposing value" which was ultimately co-created when the customer uses the firm's products and services (Vargo and Lusch, 2004). This so-called "value-in-use" is not embedded in a product or service at the moment of exchange, but rather obtained via usage processes (Tynan et al., 2010; Macdonald et al., 2009).

The dichotomous interpretation of goods and services which characterized prior research impeded a combined, simultaneous conceptualization of the two. This was finally resolved when Vargo and Lusch (2004) presented their service-dominant logic. According to Vargo and Lusch (2004), services and goods can be seen as distribution mechanisms for service provision. All economies hereby are considered as service economies (Vargo and Lusch, 2004) and all market offers can be interpreted as customer-centric product-service systems that fulfill customer needs (Ulaga and Reinartz, 2011; Shankar et al., 2009; Baines, 2007). In the following years, value-added services were integrated in the portfolio of many manufacturers to increase their customers' value-in-use.

Today, as competition quickly adapts, also (value-added) services are increasingly commoditized and no longer sufficient to guarantee a competitive advantage (Meyer and Schwager, 2007; Shaw, 2002; Schmitt, 1999). Several scholars recently identified customer experiences as the ultimate competitive element and their effective management as the most promising strategy for company's long-term success (e.g., Lax, 2012; Lemke et al. 2010; SAS, 2009; Klaus and Maklan, 2007). Pine and Gilmore (2011) even go so far to claim the rise of the Experience Economy as a logical progression from Service Economy. Addis and Holbrook (2001) interpret the overall economic development within the last decades as an evolution from mass production to mass customization, replacing supply chains with demand chains. The idea behind mass customization is to serve customers uniquely and efficiently. Hereby, mass customizing any good turns a good automatically into a service; mass customizing any service turns a service automatically into an experience (Pine and Gilmore, 2011; Addis and Holbrook, 2001). Figure 2 illustrates this Progression of Economic Value.

Figure 2: Progression of Economic Value, based on Pine and Gilmore (2011)

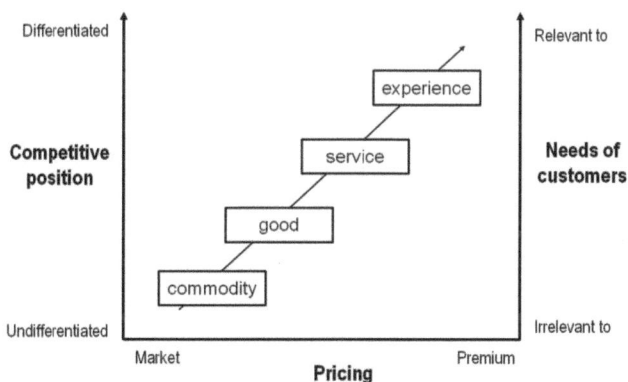

The degree of standardization continuously decreases from commodity to experience. As a company moves up the ladder it is able to fulfill individual needs better and becomes more relevant to its customers. Due to this individual focus and personal relevance, a company which stages experiences can easily differentiate from competitors and is able to charge a premium price for its offerings based on the distinctive value provided, instead of a market price dictated by competition (Pine and Gilmore, 2011).

But how can be evaluated if a company has reached the state of experience staging? How can the quality of a customer experience be measured? One measurement instrument for this purpose which gained more and more attention in the last years is the construct of 'customer experience quality'.

2.2 The Initial Conceptual Model

The conceptual model shown in Figure 3 presents the proposed relationships among the single constructs of interest: customer experience quality, perceived value, perceived wealth and customer loyalty in the form of the customer's intention to recommend the company to others and the willingness to pay more.

Figure 3 Initial Conceptual Model

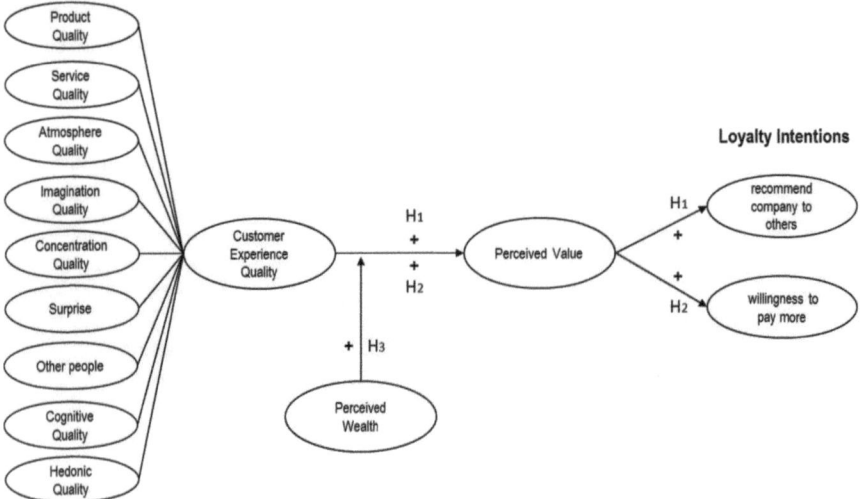

The single constructs will be subsequently described in more detail. The basic idea in a nutshell is the following: As was found in the studies of Lemke et al. (2010) and Hueiju and Wenchang (2009), perceived value is supposed to play a mediating role between customer experience quality and customer loyalty. Perceived wealth is introduced as a moderating variable. The author hypothesizes that the more wealthy customers perceive themselves, the more dimensions of (customer experience) quality they take into account when evaluating the value they receive and hereby the higher is the effect of customer experience quality on perceived value and, in turn, customer loyalty.

2.2.1 Customer Experience Quality

The existing practically oriented literature on customer experiences claims that customer experiences need to be extraordinary, memorable and compelling in order to generate a competitive advantage (e.g., Pine and Gilmore, 2011; Holbrook, 2007; LaSalle and Britton, 2003; Schmitt, 2003; Addis and Holbrook, 2001). But how is it possible to measure an experience, and moreover, how is it possible to manage customer experiences?

First of all, one needs to understand the nature of customer experiences. Customer experiences blur the traditional dichotomy of goods and services by ultimately focusing on customers' value-in-use, which is created by the orchestral combination of goods and services during the interaction between customer and company (Poullson and Kale, 2004). Customer experiences are by nature co-created by customers and lead to value perceptions both on a cognitive as well as on an affective level (Prahalad and Ramaswamy, 2004). A customer experience hereby is a holistic personal, customer specific perception of a company's overall market offering which is generated in a wide array of situations and contains a significant amount of hedonic benefits and emotional value for the customer. According to Bruhn and Hadwich (2012), the academic literature regarding customer experiences can be classified into four streams of research: product experience (e.g., Hoch, 1989), service experience (e.g., Patricio et al., 2011), brand experience (e.g., Brakus et al., 2009), and consumption experience (e.g., Hirschmann and Holbrook, 1982). The study at hand tries to merge the parallel concepts of product experience, service experience, and consumption experience by consolidating them in a holistic quality scale.

Similar to the evolution from products to services to experiences, the scales used for measuring quality have evolved and were continuously extended. Customer assessments of quality are phenomenological in nature and quality is commonly defined (e.g., Caruana et al., 2000; Zeithaml et al. 1996; Parasuramam et al. 1988) as a "perceived judgment about an entity's overall excellence or superiority" (Zeithaml, 1988, p.3). According to the evolution outlined in section 2.1, product quality can be considered as the very basic dimension of quality. It is commonly agreed on that product quality scales are embedded-value measures and have a rather limited scope of what they measure (Lemke et al., 2010; Parasumaran et al., 1988). Moreover, they are largely limited to cognitive evaluations and hereby insufficient to capture the abundance of feelings a person develops when being a customer of a specific company. Service quality scales, such as the SERVQUAL scale, (Parasumaran et al., 1988), acknowledge more quality dimensions beyond tangible product attributes (e.g. reliability, responsiveness, assurance, and empathy) but still fail to measure affective and especially hedonic components properly (Lemke et al, 2010). Therefore it is necessary to conceptualize a scale which also enables the holistic measurement of customer's feelings when experiencing to be a customer of a specific company.

A common measurement for customer experiences, although very inconsistent regarding the employed scale items, is the use of 'customer experience quality' (e.g, Maklan and Klaus, 2011; Lemke et al., 2010; Hueiju and Wenchang, 2009; Verhoef et al., 2009). Table 1 presents an overview of the most important conceptualizations of this construct, in how far the scholars related customer experience quality to other variables and which study design was used. Research context and sample size are reported as well.

In the author's opinion, Ting-Yueh and Shun-Ching (2010) so far have created the most convincing scale to measure customer experience quality. Nevertheless, they in return fail to consider product quality as a crucial element of a customer experience. Studies of Maklan and Klaus (2011) and Lemke et al. (2010) have demonstrated that customers evidentially also consider product attributes when judging customer experience quality. These studies did include product quality as a dimension of customer experience quality, but did not capture the affective components of the experience properly. Therefore, the author suggests the creation of a new scale that overcomes the pitfalls of previous research. This new scale for customer experience quality (the CXQ scale) will be developed on the basis of Ting-Yueh and Shun-Ching's (2010) five major dimensions and additionally contains the dimension product quality (see also section 3.2.2.1). Table 2 presents an overview of each of the six proposed major dimensions of the CXQ scale. Taking the sub-dimensions into account as well, customer experience quality hereby incorporates nine different dimensions. Table 2 summarizes each of the dimensions and explains in how far it affects the overall construct. The single items of the CXQ scale can be found in Table A1 and are discussed and refined in chapter three.

Table 1. Existing Conceptualizations of Customer Experience Quality

Author(s) and year	Constructs and items	Other variables in model	Study Design	n	Research context
Maklan and Klaus (2011)	Peace of mind: 6 items Outcome focus: 4 items Moments of truth: 5 items Product experience: 4 items	Loyalty Word of mouth Customer satisfaction	Online survey	218	Financial services (UK)
Lemke et al. (2010)	Communication encounter: n.a. Service encounter: n.a. Usage encounter: n.a. Experience context: n.a.	Value-in-use Relationship Outcomes	Repertory grid technique	40	B2B and B2C context
Ting-Yueh and Shun-Ching (2010)	Physical surroundings: 17 items Service providers: 5 items Other customers negative public behavior: 4 items Customers' companions: 4 items Customers themselves: 8 items	none	Interviews	146 181	Easy Shop (Taiwan) Starbucks (Taiwan)
Hueiju and Wenchang (2009)	Product Quality: 3 items Service Quality: 3 items Contextual experience: 3 items	Customer perceived value Preferences Intention	Interviews	147	Starbucks (Taiwan)
Verhoef et al. (2009)	Social Environment: n.a. Service Interface: n.a. Retail Atmosphere: n.a. Assortment: n.a. Price: n.a. Customer experiences in alternative channels: n.a. Retail brand: n.a. Customer exeriences (t-1): n.a.	Situation moderators Consumer moderators	Literature Review	0	Retailing
Gentile et al. (2007)	Sensorial/lifestlye component: 6 items Pragmatic/cognitive/lifestyle component: 4 items Pragmatic/Relational /Emotional component: 3 items	none	Mail survey	2,368	12 different B2C products

The CXQ scale includes experiential factors of which companies have direct control as well as factors they cannot control directly. Based on the ratio between controllable and uncontrollable factors, Carù and Cova (2007) introduced a "continuum of consuming experiences" that ranges from experiences which are primarily developed by the company to

Table 2. Dimensions of Customer Experience Quality

Dimension	Description
Product Quality	Lemke et al. (2010); Hueiju and Wenchang (2009)
	Products can either be tangible or intangible, but they always represent the core of the market offering and hereby also the core of the customer experience. Product Quality in this study is defined as the degree of how well the company's core product fulfills the customer's expectations of high quality, i.e. how „good" does the customer perceive the core product itself? In the case of Starbucks, product quality relates to the attributes of the tangible core product coffee. For an insurance company, the product quality relates to the (perceived) conditions of their offered insurances.
Service Quality	Lemke et al. (2010); Ting-Yueh and Shun-Ching (2010); Hueiju and Wenchang (2009); Verhoef et al. (2009)
	Besides the product, personal interaction between customers and service providers plays a crucial role for staging experiences. In this study Service Quality is defined as the degree of how well the service offered by employees fulfills customer's expectations of high quality. Particularly: are they friendly and motivated?
Store Design	Lemke et al. (2010); Ting-Yueh and Shun-Ching (2010); Verhoef et al. (2009); Gentile et al. (2007)
	When a customer interacts with a company, this can either happen physically or virtually. In the special case of Retailing, customers visit stores and interact with the physical surrounding holisticly by five senses, hereby acquiring experiences. There are three emotional sub-dimensions of experience, generated by Store Design: atmosphere, concentration, and imagination. (a) Atmosphere: customers perceive atmosphere through interaction with visual variables, kinestethic variables, olfactory variables and auditory variables. Gustatory variables are usually less involved. (b) Concentration: the store design attracts customer's attention, they enjoy themselves in the setting, are immersed in consumption and lose track of time. (c) Imagination: themed or exquisitely designed scenes stimulate customer's fantasies, activate memories, and hereby also generate experiences.
Surprise	Ting-Yueh and Shun-Ching (2010); Berman (2005); Poullsen and Kale (2004)
	Surprise (here in a positive sense) happens when cutomers receive more than they expect. This is also referred to as delight (Berman, 2005), causing sensation, excitement and memorable customer experiences.
Other people	Lemke et al. (2010); Ting-Yueh and Shun-Ching (2010); Hueiju and Wenchang (2009)
	Interactions with other people also influence customer's evaluation of experience quality. Other people can be other customers who are in the store at the same time, friends and relatives who accompany the customer or society in general.
Customer's self	Ting-Yueh and Shun-Ching (2010); Gentile et al. (2007); Poullsen and Kale (2004); Hirschman and Holbrook (1982); Holbrook and Hirschmann (1982)
	Through consumption activities, customers obtain two kinds of benefits which influence the customer's perception of the overall experience: (a) cognitive (utilitarian): learning, problem-solving, goal-orientation (b) affective (hedonic): pleasure, fun, enjoyment

experiences which are largely constructed by the consumers. For the measurement of customer experience quality this information might be negligible, however, for its management it is crucial. Verhoef et al. (2009) introduced the dimension of time to distinguish among experiences prior, at, and post purchase. Related to this differentiation, the presented CXQ scale can be seen as an aggregation across time and measures the quality of the overall customer experience with a particular focus on the phases of purchase and consumption of the product. This aggregation is considered to be sensible. As customers rather update their overall impression of a company than dissect their experiences with surgical accuracy, what is important in the end is how the customer subjectively perceives its overall quality.

2.2.2 Perceived Value

The perception of quality regarding experiences is the ultimate foundation of the proposed model but it is not sufficient to explain customer behavioral intentions. Bolton and Drew (1991), Caruana et al. (2000), Oliver (1999), and Sweeney and Soutar (2001) found that quality does not directly lead to behavioral outcomes such as purchase. Rather, it indirectly influences behavioral outcomes via a value perception that mediates the relationship. Perceived value is defined as "the consumer's objective assessment of the utility of a brand based on perceptions of what is given up for what is received" (Gupta and Zeithaml, 2006). The customer may perceive value in each phase of the customer journey, also including aspects outside the company's direct control but still related to the company, such as when interacting with other customers (Verhoef et al., 2009). On this basis the author concludes that customer experience quality as conceptualized in this research can be considered an antecedent of perceived value. Moreover, it will be hypothesized that perceived value fully mediates the relationship of customer experience quality and customer loyalty.

2.2.3 Customer Loyalty

Customer loyalty can be investigated either behaviorally or psychologically. Behaviorally, consumers are defined as loyal if they continue to buy the same product over a certain time period (Gupta and Zeithaml, 2006). This is usually measured as repeat purchase frequency or relative volume of purchasing (e.g., Tellis, 1988). However, Jacoby and Chestnut (1978) point out that this kind of behavioral customer loyalty is in some cases merely the result of convenience or high switching costs and hereby might be spurious. Customer loyalty can also be measured by psychological indicators, i.e. the customer's intention to perform a diverse set of certain behaviors. Intentions can be seen as the psychological antecedents of behavior. In terms of customer loyalty, these intentions comprise repurchase intention (e.g., Reynolds and

Arnold, 2000), intention to recommend to others, also known as word-of-mouth (WOM) (e.g., Mattila, 2001), likelihood of switching and likelihood of buying more (e.g., Selnes and Gonhaug, 2000). Zeithaml et al. (1996) merge these four aspects of loyalty into a behavioral-intentions battery with four factors – loyalty (i.e. recommend company to others), propensity to switch, willingness to pay more, and external response to service problems.

Because of the aforementioned risk of a spurious relationship between customer experience quality and behavioral customer loyalty, the author decides to apply the concept of psychological loyalty intentions. With regard to Reichheld (2003) who advocates that complex measures besides customers' intention to recommend the company to others are unnecessary to capture loyalty, the model will be kept as simple as possible. Therefore not all of Zeithaml et al.'s (1996) identified aspects will be taken into account. Only two factors are employed in the measurement model: 1) intention to recommend the company to others and 2) since people apparently are willing to accept an exponential increase in prices for experiences (Figure 1), also customer's willingness to pay more will be considered.

Hypothesis 1 and 2 connect the aforementioned constructs in the following way:

H_1 Customer experience quality affects customer's intention to recommend the company to others positively and indirectly through perceived value.

H_2 Customer experience quality affects customer's willingness to pay more positively and indirectly through perceived value.

2.2.4 Perceived Wealth

In financial economics, a consumer's wealth is defined by several observable variables, i.e. cash balances, government bonds, housing equity, stocks, other assets, and debts. This concept of wealth hereby assesses quantitatively how 'wealthy' or 'rich' a person actually is. By aggregating this wealth across a nation or a sample, and linking it to aggregated consumption in a longitudinal study, financial economists were able to find evidence for the so called 'wealth effect' which is measured by the wealth elasticity of demand (e.g., Peltonen et al., 2012; Campbell and Cocco, 2007). The wealth elasticity of demand describes the proportional change in consumption of a good relative to a change in consumer's wealth. The wealth effect predicts that an increase in wealth leads to an increase in spending, i.e. people are both willing to purchase a higher quantity of products, which stimulates repurchases, and are willing to accept higher prices.

Peltonen et al. (2012) and Campbell and Cocco (2007) find evidence for the wealth effect in several markets all over the world. However, the effect of wealth on consumption differs significantly across different countries. In order to make generalizable predictions and to be able to roll out the study globally, the author decided to analyze the respondents' wealth psychologically by measuring perceived wealth instead of measuring wealth by a set of observable financial indicators. This implies that in order to observe a wealth effect, people not actually need to be richer but merely need to perceive themselves to be richer. In the author's opinion this is a justifiable and reasonable adjustment. To the author's experience, consumers rarely are up-to-date regarding the volatile exact quantitative value of all their assets and debts but rather have a broad perception of their current financial status in mind. Measuring perceived wealth therefore can be considered a meaningful instrument when aiming to predict consumer behavior. The measurement of perception has the additional advantage that income inequalities across countries are automatically corrected. Usually these inequalities are controlled for by calculating estimates like purchasing power parities or the consumer price index. Unfortunately, as Almas and Shafir (2012) find out, even these corrected estimates are significantly biased. Perceived wealth therefore is a promising alternative for the study at hand. Since consumer perceptions are rather relative than absolute (Ariely, 2009), the respondent automatically relates his or her personal situation to others around him, hereby controlling for international income inequalities.

Jones and Mustiful (1996) investigate the differences in purchasing behavior between lower- and higher-income shoppers regarding breakfast cereals. They find out that compared to higher-income shoppers, lower-income shoppers make more rational purchase decisions as defined by consumer theory, i.e. their purchase behavior is strongly guided by their income and product prices. These findings suggest that lower-income shoppers either evaluate the quality of private label and national brands to be similar or they find the price differential to be of insufficient magnitude to justify the difference in quality. Following up on the first explanation, higher-income customers are more likely to take additional quality attributes into account when making their purchase decisions. Applied to the concept of customer experiences, less wealthy customers are supposed to have a narrower definition of quality and are supposed to care merely about the ratio of price to product quality while more wealthy customers take additional quality dimensions of the CXQ scale into account, increasing the importance of customer experience quality for perceived value.

Maslow's well known Hierarchy of Needs points towards a similar direction. Although Maslow's theory may not always hold true, it still provides a general framework for categorizing and prioritizing needs and serves as a common reasoning applied by marketers to segment markets (Schiffmann and Kanuk, 2000; Wahba and Bridwell, 1976). According to Maslow (1987), there are five layers of deficiency needs: physiological needs; safety needs; belongingness & love needs; esteem needs; and self-actualization needs. The hierarchy hereby reaches from the physical requirements for human survival at the lowest level to the desire to accomplish everything that one can in a form of mastery at the highest level. The most basic needs must be satisfied before the individual desires secondary or higher level needs and starts striving for constant betterment. Maslow and Lowery (1998) later differentiated the highest need of self-actualization in more detail and identified four subcategories: 1) cognitive: to know, to understand, and explore; 2) aesthetic: symmetry, order, and beauty; 3) self-actualization: to find self-fulfillment and realize one's potential; and 4) self-transcendence: to connect to something beyond the ego or to help others find self-fulfillment and realize their potential. For the concept of customer experiences, the cognitive and aesthetic aspect is of particular interest. Assuming that the different individual needs are increasingly satisfied with rising (perceived) wealth (Trigg, 2004), consumers who perceive themselves as more wealthy likely desire self-actualization and increasingly take cognitive and aesthetic aspects into account. As argued by Hueiju and Wenchang (2009), product quality is considered to fulfill basic needs such as physiological and security needs. Service quality represents social needs in the form of belongingness and esteem needs, whereas experience quality includes the even higher needs of self actualization, knowledge/understanding and aesthetics.

The focus of wealthy customers hereby extends from the basic dimension of product quality to higher, additional dimensions of quality, represented in the overall customer experience quality scale when evaluating the perceived value offered by a company and the impact of customer experience quality. The effect of customer experience quality on perceived value hereby increases with higher perceived wealth of the customer.

This moderating effect of perceived wealth in the model is summarized in Hypothesis 3:

H3 Perceived wealth moderates the relationship between customer experience quality and perceived value, such that the wealthier the customer perceives him- or herself, the more the perceived value is affected by customer experience quality.

3. Methodology & Research Design

In order to manage experiences in a way that customer loyalty intentions are maximized, it is crucial to understand which aspects people really remember from the purchase process. Hoch and Deighton (1989) explain that remembered purchase experiences greatly influence future behavior. That means, when deciding to choose a company, individuals first recall their past experiences. This reasoning is in line with Pine and Gilmore (1998) who claim that experiences should be memorable. Therefore, in contrast to existing studies about customer experience quality (see Table 1) in which the measurements were made right after customers/ respondents have left the store, this study will measure customer experience quality with a significant time lag. This approach ensures that only the remembered, memorable experience will be measured and linked to behavioral intentions.

The American Coffee Company Starbucks will serve as an exemplary company to which the questionnaire will be adapted in order to transfer the theory into practice and evaluate the proposed hypotheses. Starbucks was chosen because it has a worldwide network of stores with a consistent corporate design and a global strategy focused on staging experiences. Starbucks is known to many people and has already proven to be a good example in previous research on customer experiences (e.g., Chang and Horng, 2010; Hueiju and Wenchang, 2009).

As outlined in the introduction, SEM will be applied in this study. The concept of customer experience quality and the other constructs entirely based on perceptions are all latent constructs, which makes this study a perfect case to apply this advanced technique of multivariate dependence analysis. SEM enables the researcher to assess the measurement properties and test the proposed theoretical relationships in a unified and integrated manner.

The analysis of a model in SEM hereby consists of two major parts. First, the measurement model in SEM is similar to a factor analysis. On the basis of theory and/or a preliminary exploratory factor analysis which reveals the underlying structure, the researcher specifies in which way observed variables load on latent factors/constructs in the model. Afterwards, in the form of a confirmatory factor analysis, all of the equations are estimated simultaneously and it is tested whether the measurement theory holds true. Second, to test particular research hypotheses a structural model connects the latent constructs via dependence relationships. Again, the entire model is estimated at once. This procedure is similar to conducting factor analysis and a series of multiple regression analyses at once. Endogenuous constructs in the model can be compared with dependent variables in a regression, exogenuous variables are independent variables. However, this analogy has to be treated with caution. Due to the simultaneous estimation of the entire model some of the endogenuous variables might serve as a dependent variable in one relationship/equation, while being an independent variable in another relationship/equation (Malhotra, 2010).

A valuable benefit of SEM is that it explicitly takes measurement error into account, i.e. in how far do the observed variables fail to describe the latent constructs of interest.

However, in order to apply SEM appropriately, two crucial prerequisites should be fulfilled. First, SEM demands a sufficiently large sample size in order to estimate the model properly. This is the case because the χ^2-Test, which is used to decide either to accept or reject the model, significantly is influenced by sample size (n). The test is based on $\chi^2 = (n-1) \times F$, where F is the fit function between the observed sample covariance matrix and the estimated covariance matrix. Therefore, if the sample size is extremely small, the model is always accepted. In return, if n is extremely large the model is always rejected (Blunch, 2008). Recommendations about specific sample sizes fall short because the absolute required sample size depends on several characteristics of the model and cannot be generalized. As a rule of thumb, Malhotra (2010) suggests sample sizes in the range of 200 and 400 if Maximum Likelihood Estimation is used.

Second, the collected data should show multivariate normality. However, this is an assumption seldomly true in practice. Therefore, as compensation if the data deviates from the assumption of multivariate normality the sample size in return should be even larger. Regarding the effects of multivariate nonnormality, Lei and Lomax (2005) found that it does not have a significant effect on the parameter estimates but that nonnormality inflates the χ^2 statistics. This finding is in line with Henly (1993) who demonstrated that regardless of the sample size, the rejection frequency of a model is substantially higher if it is tested with a sample that contains nonnormal distributions of variables.

Taking these facts into account, the sample size for this study was planned to amount to at least 100 respondents for the exploratory factor analyses and the reliability tests in the pre-test and at least 400 respondents for the final data analysis using structural equation modeling.

In the following, the different steps of the study will be described. The theory presented in chapter two serves as the conceptual foundation as visualized in Figure 3 and as summarized in the hypotheses. The analysis will follow the classic approach of conducting SEM as suggested by Malhotra (2010).

First, in section 3.1 the constructs of interest will be assigned scales. These scales afterwards are pre-tested with an independent sample in section 3.2 where they are refined and purified with the help of exploratory factor analyses and reliability tests in SPSS. Section 3.3 summarizes the findings from the pre-test and presents the necessary adjustments concerning the scales and the overall model. The refined scales then, in section 3.4, are transferred into a measurement model with the help of the software AMOS. An additional large sample collected via an online survey is used to test both the measurement model and the structural model. The measurement model links the observed variables to the unobservable, latent constructs and allows the assessment of construct validty. Only when the measurement model can be considered valid - implying sufficient model fit and reliability - the analysis can proceed and the structural model can be specified and tested.

For more detailed introductions to the principles of SEM the interested reader is referred to Malhotra (2010), Blunch (2008) and Byrne (2001).

3.1 Assigning Scales to the Individual Constructs

In order to measure the different constructs involved in the model, the measurement instruments need to be specified. As described already in the previous chapter, existing scales for customer experience quality fail to acknowledge the full spectrum of the construct. Therefore a new scale, the CXQ scale, will be developed which aims to overcome the pitfalls of its predecessors. In order to generate a battery of items for this new scale, related scales were used as inspiration. The CXQ scale and its proposed items can be found in Table A1.

Perceived wealth requires a new scale as well because, to the knowledge of the author, it has not been measured systematically before. The items of the perceived wealth scales were entirely generated on the basis of common sense and can be found in Table A2.

To measure perceived value and loyalty intentions, already established scales were adapted to the case at hand. The perceived value scale draws all its items from the perceived value indicators scale presented by Dodds et al. (1991). The wording of the items was changed into a first person perspective to make it consistent with the rest of the questionnaire. The scale and its items can be seen in Table A3. The loyalty intentions scale basically is the behavioral-intentions scale invented by Zeithaml et al. (1996), reduced to the two dimensions of loyalty

(=intention to recommend the company to others) and pay more (=willingness to pay more). It can be found in Table A4.

All items are measured on 7 point Likert scales. An exploratory factor analysis (EFA) and a reliability test for each scale was conducted in order to 1) refine the new scales and 2) make sure the changes and reductions of the established scales had no negative effects on the usability of the scales.

3.2 Pre-Testing the Scales

A multistage development study was applied on the basis of the scale development paradigm outlined by Churchill (1979). In order to run the analyses, data was collected via an online survey. The data collection and the scale purification process are described in the following.

3.2.1 Data Collection

The data for the pre-test was collected using an online survey. The sample of in total 138 respondents was collected based on convenience by publishing the survey link on the author's facebook account. Only cases in which respondents indicated that they already have shopped at Starbucks were analyzed. After this pre-selection the final sample consisted of 123 respondents. Inspection of the 5% trimmed means for each variable showed that they did not significantly differ from the regular means (max. ±.26) and the deletion of outliers was accordingly neglected. Ages ranged from 18 to 42 years (M=22.59, SD=3.07) and the majority was German (70.7%), others were Dutch (5.7%) or of any other nationality (23.6%). 66.7% of the respondents were female, 33.3% were male.

3.2.2 Scale purification process

Following the recommendations of Churchill (1979), an iterative scale purification procedure was applied. An initial exploratory factor analysis (Principal Axis Factoring with oblique rotation and factor extraction based on Eigenvalues >1) was conducted to reveal the underlying pattern of factors. In the following, Bartlett's tests of sphericity were always significant (p<.05) and the Kaiser-Meyer-Olkin Measure of Sampling Adequacy was always >.6, indicating that the data was suitable for factor analyses.

As a second analysis tool, item-to-total correlations were taken into account. This is a common procedure when developing a scale (Kim et al., 2012). Internal consistency reliability in the form of Cronbach's alpha was considered as a third criterion to evaluate the usability of a scale.

3.2.2.1 CXQ scale

An initial exploratory factor analysis (EFA) with oblique rotation extracted six factors based on Eigenvalues >1. Some problems surfaced when analyzing this pattern matrix. The data shows an abnormal factor loading of CON1 higher than 1. In an EFA with orthogonal rotation this would indicate a severe error; however, in EFA with oblique rotation in which factor loadings are interpreted as regression coefficients instead of correlation coefficients, this is still unusual but it does not indicate a methodological error; rather it merely shows a high degree of multicollinearity and the item can be considered as highly reliable (Jöreskog, 1999). This high degree of multicollinearity in the data can be assumed to be the result of the following characteristic: since the CXQ scale measures the latent construct 'customer experience quality' with the help of several latent sub-constructs, a higher degree of multicollinearity than in usual studies with more separated factors can be assumed. This is also the reason why oblique rotation was applied.

Several items load on more than one factor, violating the assumption of unidimensionality. Hence, the items SUR1, SUR2, IMG3, AFF1, AFF2, and AFF3 are eliminated. The items OP2 and OP3 have no significant loadings on either factor and are also eliminated.

An additional factor analysis was run with the remaining items. Now only five factors were extracted. Due to the reduction from six to five factors, IMG2 loads on more than one factor and was eliminated.

A third factor analysis lead to a new pattern matrix with 15 items unidimensionally loading on five factors. Although the pattern matrix looks very promising, this scale cannot serve as the final one. The item-to-total correlations were computed for the remaining 15 items. Items which were poorly correlated ($r < .4$) to the total score were excluded; these were: COG2, PQ1 and PQ2.

A new factor analysis with the remaining 12 items was run. All items unidimensionally load on four different factors and the item-total statistics demonstrate that all the remaing items significantly correlate ($r > .4$) with the total score, which is assumed to be customer experience quality (Table 3).

Table 3. Results of Factor and Reliability Analysis for the CXQ scale

Factor/variable	Item	Factor loading	Item-to-total corr.	Cronbach's alpha
	When I think of Starbucks, I recognize that …			
Service Quality				.859
SQ1	The service is very good	.790	.495	
SQ2	The employees are friendly	1.031	.480	
SQ3	Starbucks does its best to satisfy me	.616	.611	
Atmosphere Quality				.869
ATM1	The store design is appealing	.880	.473	
ATM2	I like the design of Starbucks stores	.908	.584	
ATM3	Starbucks stores have a nice atmosphere	.595	.672	
Flow Quality				.788
OP1	I usually meet friends or relatives at Starbucks	-.461	.506	
CON1	Time passes without notice when I am at Starbucks	-.988	.602	
CON2	I often spend more time than initially intended at Starbucks	-.818	.647	
Learning Quality				.839
IMG1	The decoration of the store tells a story	-.806	.537	
COG1	I gain information about products while shopping	-.821	.714	
COG3	I learn something new while being at Starbucks	-.875	.693	

Note: Item-to-total correlation represents the correlation of the single item to the total score of the CXQ scale. Factor loadings refer to the loadings of the item on the factor under which the item is presented in the table. Cronbach's α was calculated for the single factors.

The resulting CXQ scale shows good internal consistency reliability (Cronbach's α = .869), as well do the single factors Service Quality (α = .859), Atmosphere Quality (α = .869), Learning Quality (α = .839), and Flow Quality (α = .788).

As a conclusion of the purification process it is worth to point out the implications of the findings. During the purification procedure the CXQ scale was reduced from 24 to only twelve final items and only four of the six proposed experiential dimensions were supported by the collected data. Two initially proposed dimensions of customer experience quality can immediately be recognized in the final CXQ scale: "Service Quality" and "Atmosphere Quality".

Factor three comprises two cognitive benefits items, COG1 and COG3, and the remaining imagination item IMG1 that refers to the story-telling store decoration. This shows that thoughtful decoration can help to transfer a message to the customer. According to the shared informational and educational characteristics of the items, factor three will be named "Learning Quality".

Factor four combines the two "concentration" items CON1 and CON2 with the remaining "other people" item OP1. This finding is quite logic, since meeting with friends or relatives while drinking coffee proverbially evokes that customers lose track of time, which is also referred to as a feeling of "flow" in experience literature (Csikszentmihalyi, 2008). Factor four will accordingly be named "Flow Quality".

The proposed dimension "surprise" and the other half of "customer's self", the affective benefits, are not supported. That the affective dimension is not significant as a single dimension is statistically caused by the high correlation of affective benefits items with atmosphere items. Anyway, that does not mean that experiences are merely evaluated cognitively; rather it shows that affection cannot be extracted into a single dimension because all dimensions, especially atmosphere quality, comprise affective characteristics.

Most surprising for the author is the fact that customers apparently evaluate product quality and experience quality separated from each other, since product quality could not be integrated into the CXQ scale on the basis of the collected data. However, this unexpected finding does not negatively affect the intended research.

Product quality will be integrated into the new conceptual model as a separate construct next to customer experience quality. The two product quality items form a scale which shows excellent reliability (Cronbach's α = .902) and both have a very high item-to-total correlation of .821 (Table 4).

Table 4. Results of Factor and Reliability Analysis for the Product Quality scale

Factor/ variable	Item	Factor loading	Item-to-total corr.	Cronbach's alpha
	When I think of Starbucks, I recognize that …			
Product Quality				.902
PQ1	The quality of coffee is very good	.906	.821	
PQ2	The taste of coffee is very good	.906	.821	

3.2.2.2 Perceived Wealth Scale

Factor analysis shows that all four items load on a single factor and the item-total statistics clearly indicate that each item has a high correlation (r > .4) with the total score (Table 5). Overall the scale shows acceptable internal consistency reliability with a Cronbach's α of .783. The Perceived Wealth scale therefore will be used without any purifications.

Table 5. Results of Factor and Reliability Analysis for the final Perceived Wealth scale

Factor	Item	Factor loadings	Item-to-total corr.	Cronbach's alpha
	When I think of my financial situation, I recognize that …			
Perceived Wealth				.783
PW1	I perceive myself as wealthy	.588	.511	
PW2	I can afford a good life with the money I have	.724	.633	
PW3 *	I have financial problems	.692	.580	
PW4 *	I can only afford to buy things which are really necessary	.793	.668	

Note: * indicates a negatively worded item. For the analysis the measures of this item were reversed.

3.2.2.3 Perceived Value Scale

Factor analysis shows that all three items load on a single factor and the item-total statistics clearly indicate that each item has a high correlation (r > .4) with the total score (Table 6). Overall the scale shows acceptable internal consistency reliability with a Cronbach's α of .757. The Customer's Perceived value scale therefore will be used without any purification.

Table 6. Results of Factor and Reliability Analysis for the Perceived Value scale

Factor	Item	Factor loadings	Item-to-total corr.	Cronbach's alpha
Perceived Value				.757
PV1	I consider Starbucks products to be a good buy	.523	.473	
PV2	Value for money at Starbucks is	.917	.692	
PV3	Starbucks' prices are	.740	.612	

3.2.2.4 Customer Loyalty Scale

Factor analysis shows that all six items load on a single factor and the item-total statistics clearly indicate that each item has a high correlation (r > .4) with the total score (Table 7). Based on this finding, instead of considering the loyalty intentions 'word-of-mouth' and 'willingness to pay more' as seperate, the two WPM items are considered to merge with the WOM items into a single scale which will be renamed to loyalty intentions. Overall the scale shows good internal consistency reliability with a Cronbach's α of .874. It has to be mentioned critically that the factor loading as well as the item-to-total correlation of WPM6 is rather low. Nevertheless, there will be no purifications at this stage of the analysis.

Table 7. Results of Factor and Reliability Analysis for the Loyalty Intention scale

Factor	Item	Factor loadings	Item-to-total corr.	Cronbach's alpha
Loyalty Intention				.874
WOM1	I say positive things about Starbucks to other people	.866	.782	
WOM2	I recommend Starbucks to someone who seeks my advice	.908	.804	
WOM3	I encourage friends and relatives to go to Starbucks	.905	.812	
WOM4	I consider Starbucks as my first choice to buy coffee	.811	.761	
WPM5	I continue to do business with Starbucks if their prices increase somewhat	.555	.558	
WPM6	For the benefits I currently receive at Starbucks I pay a higher price than competitors charge	.407	.406	

Note: * indicates a negatively worded item. For the analysis the measures of this item were reversed

3.3 Adjustments and Refinements

EFA showed that some of the proposed scales need further refinement. The scales for perceived wealth and perceived value remain unchanged. In contrast, the changes in the loyalty intentions scale(s) and the CXQ scale need to be transferred into a refined conceptual model (Figure 4) and new hypotheses.

The EFA for loyalty intentions extracted only a single factor instead of the two hypothesized separate factors 'intention to recommend the company to others' and 'willingness to pay more'. The effects hereby cannot be evaluated separately since loyalty intentions must be reduced to a single factor in the refined model.

Figure 4. Refined Conceptual Model after Pre-Test

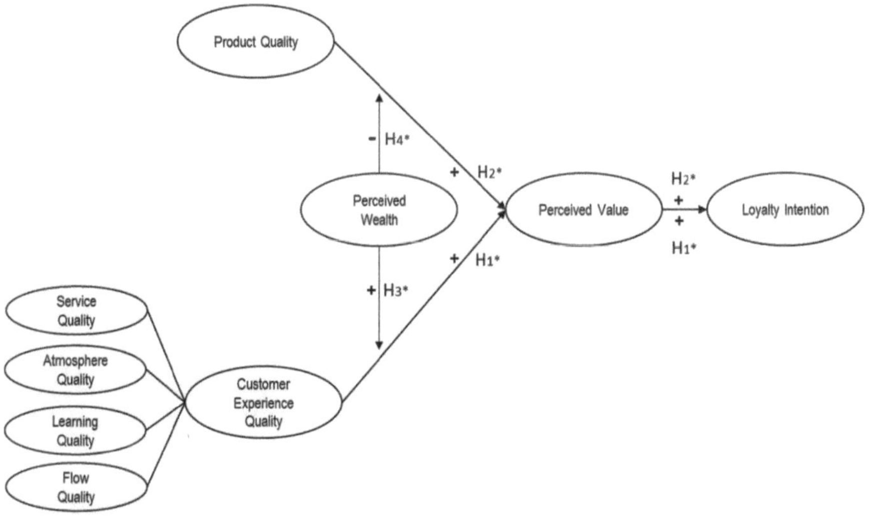

The complex CXQ scale was significantly reduced from 24 to twelve items and from nine to four first-order factors. Moreover, it was found that, in contrast to the previous argumentation in chapter 2.2, customers evaluate product quality separately from customer experience quality. Since product quality and customer experience quality are considered to be two separated constructs, the influence of product quality and customer experience quality on perceived value should be specified based on their relative importance. The reasoning of section 2.2.4 hereby becomes more radical. To account for the concept of relative importance it is hypothesized that with higher perceived wealth the focus of the customer shifts from product quality to customer experience quality when evaluating the perceived value offered by a company. That means wealthy customers do not only take additional quality dimensions into account but in return increasingly neglect the basic dimension of product quality. Hueiju and Wenchang (2009) already have proven a similar moderating effect of income on the relative importance of product quality, service quality and experience quality (see Table 1). The changes in the model in return demand the introduction of new Hypotheses to disentangle the moderating effect of perceived wealth and to respecify the mediating effect of perceived value as described above.

H_{1*}	Customer experience quality affects customer's loyalty intentions positively and indirectly through perceived value.
H_{2*}	Product quality affects customer's loyalty intentions positively and indirectly through perceived value.
H_{3*}	Perceived wealth moderates the relationship between customer experience quality and perceived value, such that the wealthier the customer perceives him- or herself, the more the perceived value is affected by customer experience quality.
H_{4*}	Perceived wealth moderates the relationship between product quality and perceived value, such that the wealthier the customer perceives him- or herself, the less the perceived value is affected by product quality.

3.4 Testing the Measurement Model

Assessing the validity of the measurement model is a necessary step before the structural model and hereby the hypotheses can be tested. A valid measurement model ensures that the scales are actually measuring what they are intended to measure, so that veritable conclusions can be drawn and valuable recommendations can be given. In the following the process for assessing and establishing a valid measurement model is described.

3.4.1 Data Collection

The data for SEM was collected using an online survey which was published in the amazon Mechanical Turk (MTurk) network. MTurk is an online crowdsourcing system which allows requesters to distribute work to a large population of anonymous workers. The tasks are limited to simple, one-time activities which are called HITs (Human Intelligence Task). For the completion of a HIT, such as answering an online survey, workers are paid a small amount of money. The completion of a survey for the study at hand, which took a worker on average 3.09 minutes, was rewarded with $0.20 for normal workers and $0.50 for so called Masters, these are workers who demonstrated great accuracy and high quality in previous tasks. The approximated hourly wage hereby amounts to $4 and $10, respectively. The collection of respondents via MTurk has three major advantages: 1) an international and diversified sample is generated that can be considered similar to Starbucks' international customer base, 2) thanks to the monetary incentive given to the respondents the data collection process is comparatively quick, and 3) by using an online survey human error is reduced as the data is transferred automatically into an SPSS data file.

The data collection process ranged from December 07 until December 25, 2012 and lead to a total sample size of 525 respondents. The text with which the link to the online survey was published in MTurk can be found in Figure A5. Equal to the pre-test, only respondents who have already shopped at Starbucks were integrated in the sample, leading to the exclusion of 26 cases. The resulting sample hereby amounted to 499 respondents. A Mardia's multivariate kurtosis value of 137.699 and a critical ratio of kurtosis of 51.845 indicates a severe violation of the assumption of multivariate normality in the collected data (treshold value for the critical ratio is 1.96). An inspection of the observations farthest from the centroid by considering the Mahalanobis d^2 statistics was used to eliminate severe outliers. The highest value amounted to 104.386 and hereby was relatively extreme. To improve normality and reduce the destructive effects of outliers, all cases with a value higher than 60 were deleted. This resulted in the exclusion of 15 cases, so that the sample had a final size of 484 respondents and a slightly improved multivariate kurtosis of 102.904 with a critical ratio of 38.158 which, nevertheless, is still much too high to assume multivariate normality.

The author considered three different approaches to deal with the violation of the multivariate normality assumption. First, the deletion of additional cases based on Mahalanobis d^2 as suggested by Gao et al. (2008) and as described above might improve the degree of multivariate normality further. However, the author decided to rather prefer a model tested on a large sample which might be affected by an inflated χ^2 statistic than a model with a lower χ^2 value which only represents an artificially reduced subsample. Second, another approach to deal with nonnormal data is known as bootstrapping and describes a resampling prodecure in which multiple subsamples of the same size as the original sample are drawn randomly with replacement. However, bootstrapping is not a panacea for nonnormal data either and does not necessarily improve the results of SEM. Bollen and Stine (1992) found that there are cases in which bootstrapping does not work and it is not possible to predict a priori when this is the case. Third, the option to adjust the data via transformations such as taking square roots or logarithms does not appear sensible either in the case at hand because it merely ensures univariate normality (Yuan et al., 2000) and makes the interpretation of the results in such a complex model extremely difficult.

Therefore, since the sample of 484 respondents is quite large and the fact that parameter estimates are not significantly biased due to nonnormal data, as found by Lei and Lomax (2005), the author decides to proceed with the analysis without any adjustments despite the violation of the assumption of multivariate normality. Nevertheless, the author is aware of the fact that nonnormal data inflates the χ^2-value and hereby leads to a higher rejection probability of the model.

The descriptive statistics show that respondents in the final study come mainly from the USA (64.5%), as well as from India (24.0%), UK (3.7%) and other countries (24.0%). Ages range from 17 to 69 years (M=32.1, SD=10.98). 47.1% of the respondents are female, 52.9% are male and the average level of education is quite high with 72.8% having at least a Bachelor's degree. The interested reader who wants to know more about the people who participate in MTurk and the representativeness of surveys conducted in MTurk is reffered to the article of Ross et al. (2010).

3.4.2 Measurement Model

The measurement model contains 58 variables of which 23 are observed and 35 are unobserved.[1] The measurement model does not distinguish between exogenous and endogenous variables. Whereas observed variables are connected with the according latent constructs via dependence relationships, all latent constructs are connected with each other via correlational/covariance relationships. In total, 91 distinct parameters need to be estimated. Perceived wealth is missing in the visualizations of Figure 5 because moderating effects can be modeled in SEM only via multi-group effects and are applied to the structural model first.

---

[1] A set of control variables (love for coffee, general attitude towards Starbucks, time of last purchase) was included in the survey as well but was excluded from the report because comparative model analyses showed that the controls do not add any value to the model and decreased the model fit significantly. Socio-demographic variables (gender, age, educational level, nationality, place of residence) were merely used as sample descriptives in order to evaluate if the sample can be considered a representation of Starbucks'customer base.

The measurement model as specified is overidentified with 220 degrees of freedom. The positive number of degrees of freedom indicates that the observed variables provide more information than necessary to estimate the 91 parameters. Model overidentification hereby allows the researcher to estimate the parameters and test the model. In general, the more degrees of freedom the more precise are the estimations and the more powerful are the tests considered to be (Blunch, 2008). Accordingly, 220 degrees of freedom are a comfortable case to start with.

3.4.3 Assessing Model Fit of the Measurement Model

Because the different measurements used to assess model fit all have their strengths and weaknesses, four different types of fit indices next to the χ^2-value as the original fit index and its according degrees of freedom will be reported: adjusted goodness-of-fit index (AGFI), root mean square error of estimation (RMSEA), comparative fit index (CFI), and the parsimony comparative fit index (PCFI).

The AGFI is a measure of absolute fit that accounts for the degrees of freedom in the model. Values of .9 and higher are considered acceptable. However, the AGFI is affected by sample size. The RMSEA is an absolute measure of badness-of-fit. Values of less than .08 are considered desirable, values between .08 and .1 indicate mediocre fit. The RMSEA adjusts the χ^2-value by taking into account the degrees of freedom and the sample size. The CFI is an incremental fit index and hereby is also useful to make comparisons across models. Values of .9 and higher are considered to indicate a good model fit (Byrne, 2001). CFI and RMSEA are the two model fit indeces which are least affected by sample size (Malhotra, 2010); however, in the presence of nonnormal data the CFI is likely to be modestly underestimated (Byrne, 2001). The PCFI additionally takes the complexity of the model into account and should be used only to compare different models as it is not suitable as an absolute measurement. The model with the higher PCFI is the better one (Malhotra, 2010; Blunch, 2008; Byrne, 2001).

When evaluating the overall fit of a model in this study, particular emphasis is put on the CFI because of its aforementioned robustness regarding sample size. When comparing two models with each other the PCFI is used as the primary indicator because it takes the potentially diverging complexity of models into account. The author hereby follows the recommendations of Hu and Bentler (1995).

Table 8 shows a summary of the model fit indices that were calculated for measurement model #1.0.

Figure 5. Measurement Model #1.0 in AMOS

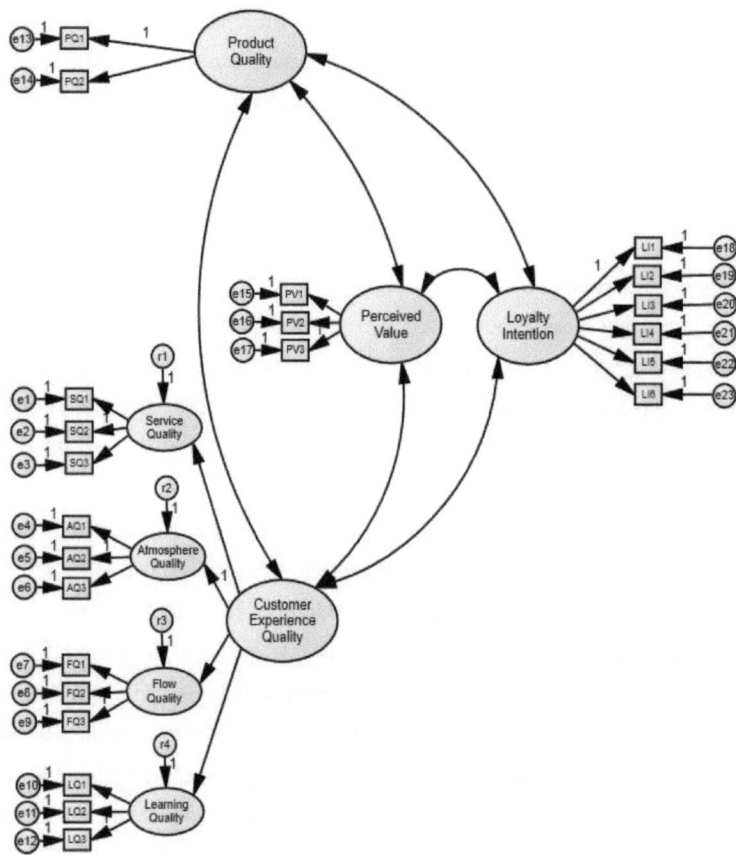

Table 8. Model Fit of Measurement Models

	model		
	# 1.0	# 1.1	# 1.2
Model Fit			
χ^2-value	1022.868	772.558	700.698
degrees of freedom	220	179	160
AGFI	.789	.809	.814
RMSEA	.087	.083	.084
CFI	.923	.937	.942
PCFI	.802	.799	.793
Misspecifications			
Number of Standardized Residual Covariances > 4	11	0	0
Number of Standardized Residual Covariances 2.5 to 4	13	3	3

The model shows a good incremental fit with the data (CFI=.923), however, when inspecting the standard residual covariance matrix, the relatively high number of values > 4 indicates that the model suffers from some severe misspecifications. Values exceeding 4.0 are considered problematic and demand the improvement of the model, those between 2.5 and 4.0 may not suggest any changes to the model but should be examined carefully (Malhotra, 2010). The absolute fit indices show a rather bad fit with the data, according to the guidelines suggested by Malhotra (2010), Byrne (2001) and Blunch (2008). Next to the high number of misspecifications, also the inflated χ^2-value of 1022.868 due to the nonnormal data is expected to contribute to this shortcoming regarding absolute fit. It seems that the observed sample covariance matrix and the estimated covariance matrix differ significantly regarding the two items SQ3 and AQ3. Therefore, these two items are deleted and a new analysis is run for the purified measurement model #1.1, whose results can be seen in Table 8.

After excluding SQ3 and AQ3 from the model, there are no severe misspecifications left that can be spotted in the standard residual covariance matrix. Also the incremental fit index was improved (e.g., CFI=.937). Surprisingly, the PCFI indicates that the previous more complex model without the purification fitted even better. However, the changes were necessary and in return improved all other fit indices, although the absolute fit indices still do not indicate a good model fit. The three remaining minor misspecifications occur between the following variables: LI5↔PV3, PQ1↔AQ1, and PQ1↔AQ2.

LI5 and PV3 both are concerned with price perceptions (see Table A1) and therefore are likely to covary more than expected by the suggested model structure. Since perceived value is supposed to lead directly to loyalty intentions, this minor misspecification is already sufficiently accounted for on the latent constructs level and can be neglected. That PQ1 covaries with AQ1 and AQ2 can be interpreted in a way that customers of Starbucks consciously or subconsciously evaluate the product quality of Starbucks coffee on dimensions not limited to the tangible attributes of the core product, but in addition extend it to the attributes of the store. The consumption environment hereby seems to play a role not only for the customer experience quality but also affects perceptions of product quality. The effect of the higher order construct customer experience quality on product quality is already integrated in the model by the correlational/covariance relationship PQ↔CXQ. Therefore, and since the standardized residual covariances have values below 4 (3.518 and 2.716 respectively) no changes of the model are undertaken.

3.4.4 Assessing Validity of the Measurement Model

Reliability of the used scales, as a necessary assumption of validity, was already proven in section 3.2 so that we can immediately proceed with the assessment of validity. By inspecting the standardized regression weights between the latent constructs and the observed variables it is possible to assess convergent validity, i.e. in how far the scale correlates positively with other measures of the same construct (Malhotra, 2006). These estimates can be interpreted as factor loadings in the form of a confirmatory factor analysis (CFA) and high values show that the observed variables load on the same construct as specified by the model. In order to establish convergent validity, all factor loadings need to be higher than .5, preferably higher than .7.

Table 9. Standardized Total Effects in Measurement Model #1.1

	PV	LI	PQ	CXQ	LQ	FQ	AQ	SQ
LQ				.888				
FQ				.840				
AQ				.622				
SQ				.562				
PV1	.906							
PV2	.916							
PV3	.878							
LI6		.490						
LI5		.771						
LI4		.839						
LI3		.923						
LI2		.940						
LI1		.943						
PQ1			.913					
PQ2			.925					
LQ1				.672	.757			
LQ2				.740	.833			
LQ3				.785	.884			
FQ1				.721		.858		
FQ2				.756		.899		
FQ3				.677		.805		
AQ1				.545			.876	
AQ2				.601			.966	
SQ1				.532				.947
SQ2				.502				.840

Table 9 shows that for measurement model #1.1, most of the loadings are higher than .7 and all of the loadings, except loading LI6 ← LI, are higher than .5. With a value of .490, the loading LI6 ← LI is slightly lower than the cutoff value and according to this variable LI6 is deleted in order to establish the necessary convergent validity of the measurement model. When taking a closer look at the CXQ scale, it has to be mentioned critically that the subdimension service quality shows a rather low loading on the second-order factor customer experience quality with a value of .562. This weakness of the scale becomes even more obvious when considering the indirect effects of SQ1 and SQ2 which are only slightly above the cutoff value of .5. Nevertheless, the author does not consider it a severe problem.

The refined measurement model #1.2 shows a CFI of .942 (Table 8), indicating a good fit with the sample data. The PCFI, however, indicates a decrease in parsimony fit taking complexity of the model into account. The resulting changes in the measurement model nevertheless were necessary and important to establish validity. As a critical comment it has to be mentioned that the absolute fit indices are still below satisfying levels. As previously discussed, this is most probably caused by the inflated χ^2-value due to the nonnormal data. Therefore, the mediocre absolute fit indices are not regarded as an enforcement to reject the measurement model. The final valid measurement model can be seen in Figure 6.

4. Data Analysis

In a structural model the latent constructs of the measurement model are connected via dependence relationships. The proposed structural model #1 (Figure 7) with a full mediation of perceived value, as suggested by the theory discussed in chapter two and three, will be estimated and described first. In order to establish validity of the proposed structural model, it will be compared with the competing conceptualizations of perceived value as a partial mediator and no mediation of perceived value at all. These nested structural models #2 and #3 are visualized in Figure 8 and Figure 9, respectively. The moderating effect of perceived wealth will be integrated in the analyses via multi-group effects. Hereby, perceived wealth is used as the grouping variable that splits the entire dataset of 484 respondents into three groups: lowPW, i.e. Likert scores of 1 to 3 (n=166); midPW, i.e. Likert scores of 4 (n=149); highPW, i.e. Likert scores of 5 to 7 (n=169). The test of the proposed theory is hereby conducted by estimating each structural model on the basis of four datasets: the full dataset with all respondents included, and three sub-datasets, including respondents characterized by low, medium, and high perceived wealth. Model fit indices were calculated to facilitate the selection of the best fitting structural model. Table 10 summarizes these measures and is used as a reference throughout the chapter.

Figure 6. Final Measurement Model #1.2 in AMOS

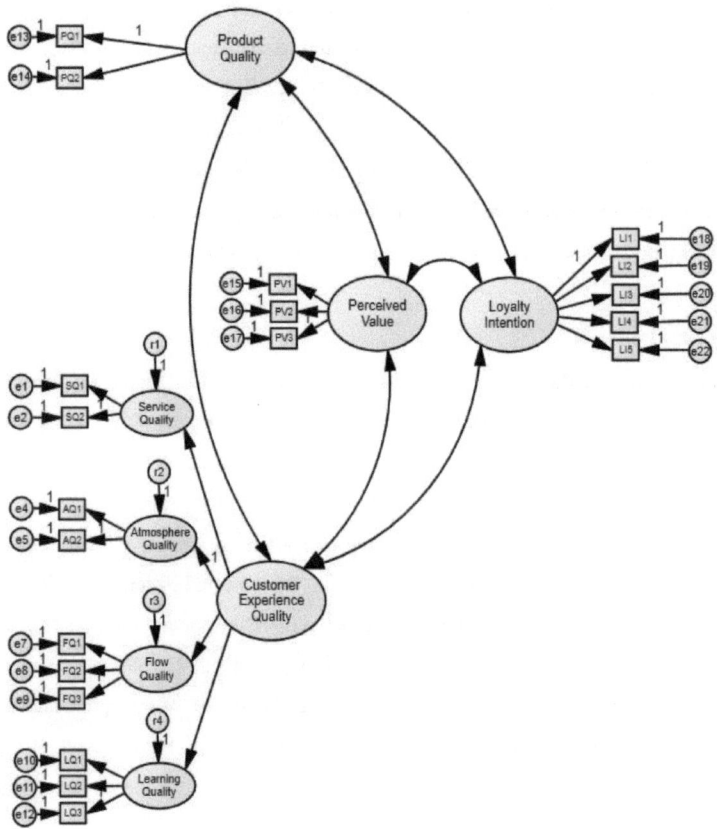

Table 10. Model Fit of Structural Models (Full Dataset)

Mediation	full	partial	none	none
Model number	# 1	# 2	# 3	# 4
Nested in Measurement Model?	yes	yes	yes	no
Model Fit				
χ^2-value	779.217	700.698	728.420	417.633
df	162	160	161	112
AGFI	.790	.814	.816	.867
RMSEA	.089	.084	.085	.075
CFI	.933	.942	.939	.958
PCFI	.796	.793	.796	.789
PRATIO	.853	.842	.847	.831

4.1 Comparison of Competing Models

The standard approach to decide among models is to test if they differ significantly in terms of their differences in χ^2. Regarding the absolute χ^2-values, each of the three models by far exceeds the critical value of 192.700, 190.516, and 191.608 respectively (at a significance level of 0.05) and formally must be rejected. However, the χ^2 - as in this case - is not a recommendable fit index in many situations in practice because it is affected by three major factors: 1) sample size, 2) model size, and 3) distribution of variables (Tanaka, 1993). This means, the larger the sample size and the more variables are in the model and the less the sample data shows multivariate normality, the higher is the χ^2-value and the more likely the model is rejected. In the case at hand, the model complexity is significantly increased by the second-order construct structure of the CXQ scale and, moreover, the sample data shows severe multivariate nonnormality.

As was expected regarding the findings in the measurement model, the inflated χ^2 leads to a mediocre performance of all structural models in terms of absolute fit. Since the structural models #1, #2, and #3 all are nested in the measurement model, they cannot have a lower χ^2-value (Blunch, 2008). As the χ^2-value additionally is the basis for all fit indices, the outcome that none of the models achieves satisfying levels of AGFI or RMSEA values is not surprising. The author therefore does not take the indication of the χ^2 and the absolute fit indices as an enforcement to reject any of the models per se. Instead, other fit indices will be taken into account. The relatively robust incremental fit index CFI indicates a good fit for all three models. On one hand this might be pleasing, on the other hand this makes the decision which model to choose less clear. In order to make a proper decision which model to choose best, the author relies on a comparison based on the parsimony fit index as primary criterion and incremental fit as second criterion.

Although researchers should evaluate model fit of a single model independent of parsimony considerations, parsimony is a crucial aspect when comparing alternative theories of different complexity. That means, if simpler alternative models seem to be as good in predicting outcomes, the researcher should favor the simpler model (Malhotra, 2010).

The parsimony ratio (PRATIO) is the basis for most of the parsimony fit indices and represents the number of constraints in the model being evaluated as a fraction of the number of constraints in the independence model (Mulaik et al., 1989; James et al., 1982). A multiplication of the PRATIO with the CFI yields the PCFI.

When inspecting Table 10, we can see that the incremental fit (CFI) for structural model #2 and #3 are better than for #1 (CFI_2=.942 and CFI_3=.939 vs. CFI_1=.933). However, when taking into account parsimony (PCFI), the proposed structural model #1 outperforms #2 and is on one level with #3. The results of the model comparison hereby are quite ambiguous.

Overall, the author relies more on the parsimony fit indeces which favor structural model #1 and #3 but does not refuse competing model #2 yet. Instead, as a response to the ambiguous findings, the author continues to analyze all structural models in terms of their implications regarding the relationships among the constructs of interest first.

During this process, the author faced a so called Heywood case in each of the models for the low perceived wealth group, i.e. the variance of error term e5 (which belongs to AQ2) was negative - a case that is theoretically not possible but that not seldomly occurs in SEM.

The presence of a Heywood case leads to an inadmissable solution since the other estimates cannot be considered reliable anymore (Rindskopf, 1984). A common cause of Heywood cases is a failure to represent each latent construct with a sufficient number of observed variables with large loadings. McDonald (1985) suggests that every latent construct should be defined by at least three, and preferably four, observed variables with large loadings on it. In the model at hand there are three constructs (PQ, SQ, and AQ) which are represented by only two observed variables, because the process of establishing validity for the measurement model reduced their previously sufficient number. However, it is possible to work around Heywood cases by placing additional constraints which are not part of the proposed theory but which are also not against it (McDonald, 1985). As the models are multiple-group models and the Heywood case did not appear in the analysis for the full dataset, the author decided to constrain the variance of error term e5 for each model to be equal across groups. Accordingly, the variance of e5 was fixed for all groups to the value which was estimated for the full dataset, i.e. .064 for #1; .065 for #2; .068 for #3. Hereby the Heywood cases were removed and the additional constraint in each model only marginally affected the other estimates (Table A6, Table A7, Table A8).

4.1.1 Interpretation of Structural Model #1 suggesting Full Mediation

Despite the mixed findings of the model fit indices, the author continued the analysis and examined the particular parameter estimates to see if the hypotheses are supported. Table 11 summarizes the standardized regression weights for the dependence relationships as well as the correlations between product quality and customer experience quality in structural model #1 for the entire dataset and the three distinct perceived wealth groups.

Figure 7. Structural Model #1

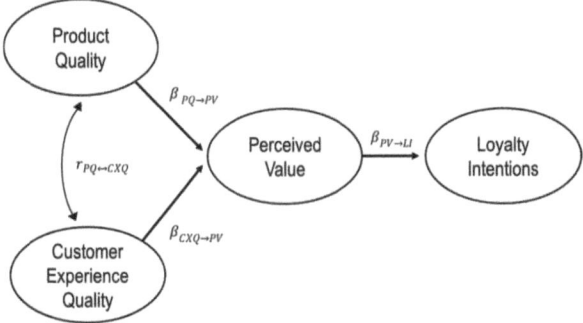

The conception of perceived value as a full mediator seems to make sense. Data shows that product quality and customer experience quality both significantly affect perceived value ($\beta_{allPW,PQ \rightarrow PV}$=.237, p=.000; $\beta_{allPW,CXQ \rightarrow PV}$=.704, p=.000) and perceived value significantly affects the customer's loyalty intentions ($\beta_{allPW,PV \rightarrow LI}$=.884, p=.000). In the different perceived wealth groups this pattern is always replicated except for the group of high perceived wealth, where $\beta_{highPW,PQ \rightarrow PV}$ is highly insignificant (p=.935). Nevertheless, this alleged inconsistency is not contradicting and will be explained in the following.

Table 11. Multi-group effects of Perceived Wealth in Structural Model #1 (e5 fixed)

Multi-group effects	allPW	lowPW	midPW	highPW	effect PW	support
$\beta_{PQ \rightarrow PV}$	.237	.444	.202	(-010)	-	H2*, H4*
$\beta_{CXQ \rightarrow PV}$	.704	.445	.817	.914	+	H1*, H3*
$\beta_{PV \rightarrow LI}$	.884	.808	.893	.916	+	H1*, H2*
$r_{PQ \leftrightarrow CXQ}$	.699	.603	.610	.837	+	

Note: Heywood case has been removed by constraining the variance of e5 to be equal across groups (0.064)
( ) insignificant effect for p=0.05

When inspecting the development of the standardized regression weights across the different perceived wealth groups, one can identify certain trends. As hypothesized by H4*, the effect of product quality on perceived value diminishes with increasing perceived wealth. This moderating effect of perceived wealth makes the PQ→PV effect for the high perceived wealth group even so small that it becomes insignificant. The data hereby supports H4*, i.e. perceived wealth has a negative moderating effect on the relationship between product quality and perceived value. Hypotheses H1* and H2*, suggesting perceived value as a full mediator in the model, are supported as well.

The regression weights additionally show that customer experience quality overall and in each group has a higher effect on perceived value than product quality. The inclusion of customer experience quality as represented by the CXQ scale hereby significantly increases the researcher's ability to predict customer loyalty intentions.

As hypothesized by H3*, perceived wealth has a positive moderating effect on the relationship between customer experience quality and perceived value. The data shows that $\beta_{CXQ \to PV}$ continuously increases the higher the perceived wealth of the customer gets and is hereby in support of H3*.

The fact that product quality has no significant effect on perceived value for these customers might be misleading though and one might conclude that product quality is not important for these customers. The author, however, interprets the result in the way that for wealthy customers, product quality loses its role as a differentiation criterion and fades into a qualification criterion regarding value generation. Hereby, product quality is a necessary but not sufficient condition to create value for the customer with high perceived wealth. As can be seen across the different groups, $r_{PQ \leftrightarrow CXQ}$ increases with the perceived wealth of the customer. This means, the wealthier the customer perceives him- or herself, the more the two major quality dimensions merge and the more they are evaluated jointly. This supports the previously described theory that product quality is still important but took over the role of a mere qualification criterion.

To conclude, regardless the reluctant model fit indices, all hypotheses as suggested by the theory discussed in chapter three are supported by the data.

4.1.2 Interpretation of Structural Model #2 suggesting Partial Mediation

As Table 12 shows, perceived value has a significant effect on loyalty intentions ($\beta_{allPW,PV \to LI}$=.383, p=.000). Moreover, product quality and customer experience quality both significantly affect perceived value ($\beta_{allPW,PQ \to PV}$=.200, p=.000; $\beta_{allPW,CXQ \to PV}$=.696, p=.000) and loyalty intentions ($\beta_{allPW,PQ \to LI}$=.175, p=.000; $\beta_{allPW,CXQ \to LI}$=.415, p=.000). One could deduce that perceived value has the role of a partial mediator. This pattern, however, merely holds true when excluding the moderating effects of perceived wealth. Indeed, when estimating the parameters for the different groups of perceived wealth, the pattern is very inconsistent and changes arbitrarily with several effects becoming insignificant. A consistent moderating effect of perceived wealth hereby cannot be identified and there is no consistent partial mediation through perceived value.

Figure 8. Structural Model #2

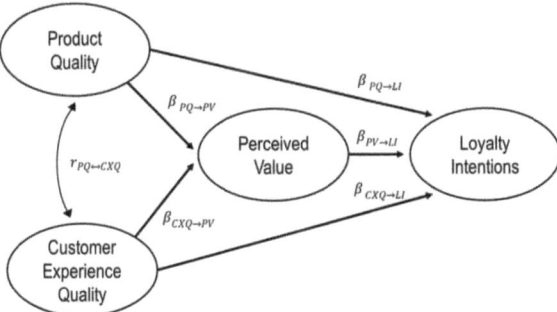

The diverging effects across groups are the following: For low perceived wealth, product quality merely has an indirect effect through perceived value, whereas customer experience quality has both a direct and indirect effect. For medium perceived wealth, product quality and customer experience quality both only have a direct effect. For high perceived wealth, product quality only has a direct effect whereas customer experience quality only has an indirect effect.

Table 12. Multi-group effects of Perceived Wealth in Structural Model #2 (e5 fixed)

Multi-group effects	allPW	lowPW	midPW	highPW	effect PW	support
$\beta_{PQ \to PV}$	.200	.464	(.105)	(-.070)	-	[H2*], H4*
$\beta_{CXQ \to PV}$	.696	.374	.841	.941	+	[H1*], H3*
$\beta_{PV \to LI}$	.383	.468	(.072)	.494	?	
$\beta_{PQ \to LI}$	.175	(-.024)	.362	.211	?	
$\beta_{CXQ \to LI}$	.415	.494	.587	(.285)	?	
$r_{PQ \leftrightarrow CXQ}$	.706	.623	.612	.837	?	

Note: Heywood case has been removed by constraining the variance of e5 to be equal across groups (0.065)
( ) insignificant effect for p=0.05 ; [ ] Hypothesis is partially supported

Thus, H1* and H2*, suggesting an indirect effect of product quality and customer experience quality on loyalty intentions, are partly supported by the data from the full dataset but need to be rejected when introducing perceived wealth as a moderator. H3* is supported as $\beta_{CXQ \to PV}$ increases with perceived wealth. H4* might be accepted based on the reasoning presented in

4.2.1, i.e. the moderating effect of perceived wealth is so strong that the effect described by $\beta_{PQ \rightarrow PV}$ gets so small that it becomes insignificant. However, H3* and H4* alone, without a significant and consistent effect of perceived value on loyalty intentions, have only limited explanatory power.

4.1.3 Interpretation of Structural Model #3 suggesting No Mediation

The theory of having no mediation of perceived value at all implies that product quality and customer experience quality have only direct effects on loyalty intentions. The estimates for this model can be seen in Table 13. Since perceived value was excluded as a mediator and is supposed to only correlate with product quality and customer experience quality, all hypotheses are rejected per se. When interpreting the estimates for the full dataset, excluding perceived wealth as a moderator, both product quality and customer experience quality have significant direct effects on loyalty intentions ($\beta_{allPW, PQ \rightarrow LI}$=.133, p=.000; $\beta_{allPW, CXQ \rightarrow LI}$=.815, p=.000).

Figure 9. Structural Model #3

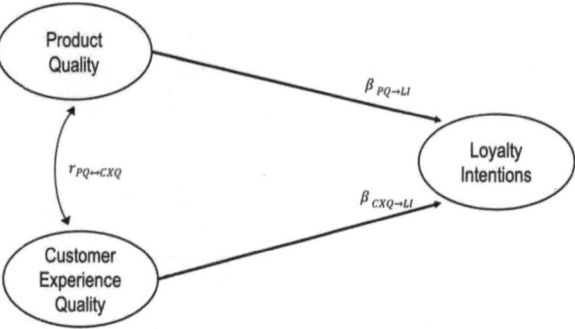

As an interesting finding that still is somewhat in line with the theory proposed in chapter two, it can be pointed out that when considering the direct effects only, for all groups, customer experience quality has a way higher effect on loyalty intentions than has product quality. Despite that, when analyzing the development $\beta_{PQ \rightarrow LI}$ and $\beta_{CXQ \rightarrow LI}$ across groups, no consistent trend in the form of a linear moderating effect of perceived wealth can be identified for the dependence relationships.

When inspecting the differences across the perceived wealth groups, one cannot identify certain patterns in the form of a consistent moderating effect. According to the data, product quality only has a significant effect on loyalty intentions for the medium perceived wealth group.

Table 13. Multi-group effects of Perceived Wealth in Structural Model #3 (e5 fixed)

Multi-group effects	allPW	lowPW	midPW	highPW	effect PW	support
$\beta_{PQ \rightarrow LI}$	.133	(.023)	.363	(-.107)	?	
$\beta_{CXQ \rightarrow LI}$	.815	.872	.657	1.040	?	
$r_{PQ \leftrightarrow CXQ}$	.735	.677	.614	.875	?	
$r_{PQ \leftrightarrow PV}$	.691	.693	.621	.718	?	
$r_{CXQ \leftrightarrow PV}$	.894	.798	.911	.928	+	

Note: Heywood case has been removed by constraining the variance of e5 to be equal across groups (0.068)
( ) insignificant effect for p=0.05

4.2 Selection of the Best Fitting Structural Model

Previous analysis showed that the selection of one of the structural models as the best fitting one is not obvious, but rather ambiguous. As previously stated, the author relies on the incremental and parsimony fit indices to make the decision.

Although the incremental fit is highest for structural model #2, (CFI_2=.942 vs. CFI_1=.933 and CFI_3=.939), the model of partial mediation is outperformed by the models #1 and #3 when taking into account parsimony ($PCFI_2$=.793 vs. $PCFI_1$=$PCFI_3$=.796).

As previously explained, the author considers parsimony a crucial component when comparing models of diverging complexity and accordingly puts the highest importance on the PCFI. Structural model #1 and #3 hereby are equally good, contrasting model fit and complexity, while structural model #2 is worse. The final decision boils down to two extreme cases: either support the theory of perceived value as a full mediator or assume no mediation of perceived value at all. Furthermore, the selection of structural model #1 would lead to the support of all hypotheses whereas the selection of structural model #3 would lead to the entire opposite, i.e. the rejection of all hypotheses.

Following up on the theory of no mediation, it implies that perceived value at least in future studies would be entirely excluded from the questionnaire/model. This line of reasoning was continued and a structural model #4 was estimated in which the construct of perceived value was entirely omitted. It has to be pointed out that this structural model #4 is neither nested in the previously discussed measurement model nor the other structural models. Nevertheless, a discussion of the according model fit indices for this competing model finally allows a justifiable decision for one of the models. Although the incremental fit of structural model #4 is higher than for #1 (CFI_4=0.958 vs. CFI_1=0.933), the parsimony fit measure in contrast clearly prefers structural model #1 ($PCFI_1$=0.796 vs. $PCFI_4$=0.789). In the author's opinion, the parsimony fit index should serve as the ultimate decision criterion when comparing two models of diverging complexity. Based on these findings and in accord with the theory presented in chapter two, the author considers 1) perceived value a full mediator in the relationships PQ→PV→LI and CXQ→PV→LI, and 2) perceived wealth a moderator of the relationships PQ→PV and CXQ→PV.

5. Discussion

The different model fit indices gave contradicting recommendations which structural model finally to choose. The author in his decision to use the PCFI as the primary evaluation criterion followed the recommendations of Hu and Bentler (1995) and took the critical factors into account, i.e. sample size, model size, estimation procedure, misspecifications and multivariate normality. Regardless which fit indices are finally chosen as evaluation criteria, Byrne (2001) cautions that global fit indices alone cannot serve as the singular judgement to determine a model's adequate fit with the sample data.

It certainly is a shortcoming of the selected structural model that the absolute fit indices merely indicate mediocre absolute fit. The author refers to sources such as Gao et al. (2008), Lei and Lomax (2005), Byrne (2001), Hu and Bentler (1995) and believes that the multivariate nonnormality, while not biasing the parameter estimates, had a significant negative impact on the (absolute) fit indices. The author's decision not to take the mediocre absolute fit indices as an indication to reject the model is supported by Sobel and Bohrnstedt (1985, p. 158) who argue that "scientific progress could be impeded if fit coefficients (even appropriate ones) are used as the primary criterion for judging the adequacy of a model." Therefore, the fact that structural model #1, chosen due to its superiority in parsimony fit, is in line with previous research findings, strengthens the credibility of the decision.

The ambiguity of the findings though may serve as an explanation for the diverging conceptualizations in the literature concerning customer experience quality. The trust of many researchers in absolute and incremental model fit indices as part of the confirmatory nature of SEM may have encouraged them to accept explanations for phenomena which could have been explained better by a competing theory. Therefore, the comparison of competing models, although tedious, is a crucial step to validate a theory. In the case at hand the theory of a full mediation cannot be considered an exclusive explanation for the observed phenomenon, but is considered to be the best one.

6. Conclusion

The study at hand had the aim to explain the success of companies who stage experiences in today's markets. Moreover, the results should enable managers to stage experiences more effectively and efficiently. This practical aim was translated into three major goals from a theoretical perspective: 1) develop a holistic measurement scale for customer experience quality, 2) link customer experience quality to customer loyalty, and 3) investigate if the perceived wealth of customers has a moderating effect, so that customers characterized by diverging levels of perceived wealth differ in their perceptions and demands.

All these goals are achieved.

6.1 Theoretical Implications

The newly developed CXQ scale measures customer experience quality on the basis of four distinct factors: service quality, atmosphere quality, flow quality, and learning quality. Thanks to the extensive analysis in the form of EFA, Reliability Tests, and CFA, the scale can be considered reliable and valid. EFA yielded the surprising but essential finding that product quality is not a sub-dimension of customer experience quality. Accordingly, the two constructs were modeled separately in the further analysis. For the CXQ scale, the initial battery of 24 items was reduced to ten items (see 6.3). In addition to the CXQ scale, a new scale for measuring perceived wealth was invented that demonstrated reliability and validity as well. The perceived wealth scale kept its initial four items of which two are negatively worded. The scale development process followed the recommendations of Churchill (1979) for the EFA and additionally applied a CFA as suggested by Malhotra (2010).

On the basis of previous research findings in the fields of marketing, psychology and financial economics, a conceptual model was proposed that gave an initial explanantion in which way customer experience quality affects customer loyalty intentions. The author suggested a full mediation through perceived value. Moreover, perceived wealth was introduced as a moderator that strengthens the effect of customer experience quality on perceived value, whereas it diminishes the effect of product quality on perceived value. With the help of SEM and a set of competing structural models, which suggested partial and no mediation, the theory of a full mediation through perceived value and perceived wealth as a moderator was entirely supported. The findings hereby yield two essential insights. First, customer experience quality per se has a much higher effect on perceived value and, in succession, customer loyalty intentions than has product quality. Hence, the establishment of high customer experience quality is a successful strategy to create loyal customers in today's markets. Second, for customers who perceive themselves as more wealthy, the importance of customer experience quality is highest, whereas product quality even has no significant effect on perceived value. However, this finding might be misleading and the author points out that product quality is still important for customers with high perceived wealth. For wealthy customers though, product quality rather loses its role as a differentiation criterion and fades into a qualification criterion. For these customers, providing product quality is merely a necessary but not sufficient condition to generate value and customer loyalty intentions. Perceived wealth of respondents can hereby serve as a valuable segmentation variable for marketers and should be included in future research at least as control variable.

Furthermore, the study yieled an additional finding that explains the co-existence of diverging conceptualizations of customer experience quality. Regarding the measurement of customer experience quality, an undecided debate among researchers is the question to include or exclude product quality as a sub-dimension (e.g., Maklan and Klaus, 2011; Ting-Yueh and Shun-Ching, 2010; Hueiju and Wenchang, 2009). Although EFA clearly demonstrated that product quality cannot be considered a sub-dimension of customer experience quality, the analysis via SEM revealed that with rising perceived wealth of the customer, product quality and customer experience quality are increasingly evaluated jointly, indicated by a rising correlation between the two constructs. This occurrence might have encouraged other researchers to consider product quality a sub-dimension of customer experience quality.

6.2 Managerial Implications

The thesis at hand revealed and proved the psychometric pattern which explains why and how staging experiences leads to long-term success in today's markets. The findings of this study bring companies one major step forward towards staging experiences which are individually tailored to diverging customer segments and which hereby effectively and efficiently trigger desirable customer behavior. With regard to evaluating a company's performance, the essential basis of all companies' strategies and actions boils down to two clear goals: maximize profits and minimize risk (Nakada, 2005). Accordingly, the important question when advising managers is: in how far does this study contain information that helps managers to increase profits and lower risk for the company? This section will give answers and recommendations to this topic.

Customer loyalty is an omnipresent keyword in company's board meetings and marketing theory. Customer loyalty is crucial for a company because it both affects risk and profits. Pauwels and Reibstein (2010) found that high customer loyalty lowers risk levels of sales, as loyal customers buy products not only repeatedly but also more predictably because they cover their demand for a specific product at one single company/store instead of shopping around. Moreover, loyal customers serve as unpaid advocates and advertise the company without being paid for. Loyal customers additionally are willing to pay higher prices. Both word-of-mouth and higher prices increase revenues. This study has proven that excellent customer experiences evoke customer loyalty intentions and hereby lay the basis for the establishment of actual customer loyalty in the form of repurchase, cross-buying, lower price elasticity, and word-of-mouth.

Whether customer loyalty results in an increase of profits in the end is dependent on the costs for establishing it (Rust et al., 2002). The creation of excellent experiences is usually linked to high investments in facilities, training of employees, market research, and constant improvement. Therefore, the manager needs to consider consciously whether such high investments pay off in the end. As successful examples in practice (e.g., Starbucks, McDonald's, Dell, Luxory Hotels) show, experience staging can serve as a profitable strategy that ensures company's long-term success.

The newly developed CXQ scale hereby facilitates the management of experiences as it is a handy, aggregated measure of the overall customer experience while simultaneously allowing the seperate analysis of its four sub-dimensions. Thanks to these characteristics, managers are able to detect precisely *what* works well or poorly in the company-customer interactions instead of merely knowing *that* something works well or poorly somewhere. Data which is collected via the CXQ scale can serve as an analysis tool which gives the manager a profound overview of customers' perceptions of the company and the opportunity to initiate purposive strategies to improve single facets of the customer experience. It has to be considered that in the attempt to achieve excellence, the manager can only maximize the overall customer experience quality by maximizing each of the four quality sub-dimensions.

In case the manager decides not to lead the company to the next level of customer value generation, he or she has to keep in mind that product quality has lost its role as a differentiation criterion and merely is a qualification criterion in today's business-to-consumer markets. Selling products with superior characteristics is not sufficient anymore to generate loyal customers who spread the word and are willing to pay a premium price. In contrast, customer experience quality can serve as a powerful predictor of customer loyalty and accordingly should find its way into each company's marketing strategy.

6.3 Limitations & Future Research

Although all scales in this study demonstrated reliability and validity, the CXQ scale contains some room for improvement. Since the factors service quality and atmosphere quality are only represented by two items, problems might occur during further analyses. The addition of supplementary items to the single factors would increase the scale's power further and make it a very useful tool both for theory and practice. In general, when developing new scales or using established scales for SEM, researchers should keep in mind that constructs should be represented by at least three, preferably four, items to contain a buffer for the potential deletion during further analyses (MacDonald, 1985). Although the single factors of the new CXQ scale were initially represented by a sufficiently large number of three observable variables/items, the deletion of items during the validation process of the measurement model lead to inconveniences in analyses of the structural models, i.e. presence of Heywood cases in the low perceived wealth groups. The author worked around the negative variances in error terms by setting an additional constraint and the Heywood cases had no negative effect on the interpretability of the results; however, they can easily be avoided by keeping the recommendations mentioned above in mind.

Moreover, the newly developed CXQ scale does not explicitly distinguish between purchase and consumption experience. Rather it implies that both events more or less occur together. In the exemplary case of Starbucks, purchase and usage happen nearly at the same time and at the same spot. This simplification was intended to facilitate research but is a restriction for future studies if the scale is intended to be used without further adjustments.

However, the CXQ scale and the findings of this study can be extended to most services in general. The inseperability of delivery and consumption of services is commonly acknowledged (Hartmann and Lindgren, 1993) and services hereby do not conflict with the scale's restriction in any way. This generalization is in line with the argumentation of Addis and Holbrook (2001) who say that experiences merely are services which are mass-customized. What is important to keep in mind as the necessary assumption to apply the CXQ scale is that all quality dimensions should be attributable to a single company. Hence, the application of the CXQ scale to the area of classical retailing is not recommendable. Besides the fact that purchase and consumption experience usually diverge, perceptions of customers towards different companies, i.e. retailers and manufacturers, would be mixed.

Moreover, the study discussed customer experiences in a strict offline setting and findings should not be generalized to online shopping. Besides the improper wording of the CXQ scale items, the purchase behavior of online shoppers might significantly differ, especially regarding price sensitivity (Junhong et al., 2008).

Experiences are by nature a phenomenon that subjectively affects individual consumers during the interaction with a company. The study at hand hereby took a strict business-to-consumer marketing perspective. Although it is argued that organizational learning is based on individual learning (Song et al., 2008), the insights from this study are not generalizable for business-to-business markets. When choosing a supplier, a company/manager rather relies on rational decisions and is more affected by cognitive than affective stimuli. Experiences at least as conceptualized in this study are supposed to have only a small influence on decisions and behavior in these situations.

The author suggests the analysis with an additional sample, collected for another company. Hereby the generalizability of the model can be examined. Other companies in the hospitality business (e.g. cafés, restaurants, bars) probably will not differ significantly. A valuable study though might be to test in how far the model can be applied to (flagship) stores of a single manufacturer. Stores of the American consumer electronics company Apple for example also are known to provide an experience for customers. It would be interesting to see if product quality has a higher impact on perceived value in this case, considering that the products are far more expensive than a cup of coffee and that the customer uses the product beyond the time of the purchase experience. Following up on this reasoning, an additional moderator such as 'price of the product' or 'durable vs. non-durable good' could be integrated in the model that moderates the effect of product quality on perceived value.

Moreover, future studies should investigate in which way customer intentions actually lead to the proposed customer behavior and business outcomes by including "hard data" in the form of number of repurchases, price paid by customers, store revenue or financial data in general. Since customer experiences are usually costly to stage, these insights could quantitatively prove whether customer experience quality does only generate higher value for the customer or in return also increases the profitability of the firm. The latter can merely be assumed in this study, based on the argumentation outlined in section 6.2 and depends on each company's individual situation.

Reference List

Addis, M. & Holbrook, M.B. (2001). On the conceptual link between mass customisation and experiential consumption: An explosion of subjectivity. *Journal of Consumer Behaviour*, 1 (1), 50–66.

Almås, I. (2012). International Income Inequality: Measuring PPP Bias by Estimating Engel Curves for Food. *American Economic Review*, 102(2), 1093-1117. doi:10.1257/aer.102.2.1093

Ariely, D. (2009). Predictably Irrational. The hidden forces that shape our decisions. Harper Collins, New York.

Baines, T.S. (2007). State-of-the-art in product-service systems. Proceedings of the Institution of Mechanical Engineers – Part B – Engineering Manufacture, 221, 10, pp. 1543–1552.

Berman, B. (2005). How to Delight Your Customers. *California Management Review*, 48(1), 129-151.

Blunch, N. (2008). Introduction to Structural Equation modeling using SPSS and AMOS. Sage, London.

Bollen, K. A., & Stine, R. A. (1992). Bootstrapping Goodness-of-Fit Measures in Structural Equation Models. *Sociological Methods & Research*, 21(2), 205.

Bolton, R. N., & Drew, J. H. (1991). A multistage model of customers' assessments of service quality and value. *Journal of Consumer Research*, 17, 375–384.

Brakus, J., Schmitt, B., & Zarantonello, L. (2009). Brand Experience: What Is It? How Is It Measured? Does It Affect Loyalty?. *Journal Of Marketing*, 73(3), 52-68. doi:10.1509/jmkg.73.3.52

Bruhn, M. & Hadwich, K. (2012). Customer Experience: Forum Dienstleistungsmanagement. Gabler, Wiesbaden.

Byrne, B. (2001). Structural Equation Modeling with AMOS: Basic Concepts, Applications, and Programming. Lawrence Erlbaum Associates, London.

Campbell, J. Y., & Cocco, J. F. (2007). How do house prices affect consumption? Evidence from micro data. *Journal Of Monetary Economics*, 54(3), 591-621. doi:10.1016/j.jmoneco.2005.10.016

Carù, A. & Cova, B. (2003). Revisiting consumption experience: A more humble but complete view of the concept. *Marketing Theory*, 3 (2), 267–286.

Carù, A. & Cova, B. (2007). Consuming experience. Routledge, London.

Caruana, A., Money, A. H., & Berthon, P. R. (2000). Service quality and satisfaction: the moderating role of value. *European Journal of Marketing*, 34(11/12), 1338–1353.

Christopher, M. (1996). From brand values to customer value. *Journal of Marketing Practice: Applied Marketing Science*, 2 (1), 55-66.

Churchill Jr., G. A. (1979). A Paradigm for Developing Better Measures of Marketing Constructs. *Journal Of Marketing Research* (JMR), 16(1), 64-73.

Copeland, M. T. (1923). RELATION OF CONSUMERS' BUYING HABITS TO MARKETING METHODS. *Harvard Business Review*, 1(3), 282-289.

Csikszentmihalyi, M. (2008). Flow: The Psychology of Optimal Experience. HarperPerennial, New York.

Dodds, W. B., Monroe, K. B., & Grewal, D. (1991). Effects of Price, Brand, and Store Information on Buyers' Product Evaluations. *Journal Of Marketing Research* (JMR), 28(3), 307-319.

Euromonitor International (2008). Live it up! In the Americas, experience is everything for consumers.

Gao, S., Mokhtarian, P. L. & Johnston, R. (2008). Nonnormality of Data in Structural Equation Models. *Journal of the Transportation Research Board*, 2082, 116-124.

Gentile, C., Spiller, N., & Noci, G. (2007). How to Sustain the Customer Experience:: An Overview of Experience Components that Co-create Value With the Customer. *European Management Journal*, 25(5), 395-410. doi:10.1016/j.emj.2007.08.005

Grönroos, C. (1997). From marketing mix to relationship marketing - towards a paradigm shift in marketing. *Management Decision*, 35(3/4), 322.

Gupta, S., & Zeithaml, V. (2006). Customer Metrics and Their Impact on Financial Performance. *Marketing Science*, 25(6), 718-739.

Hair, J. F., Tatham, R. L., Anderson, R. E., & Black, W. C. (2006). Multivariate data analysis (6th ed.). Prentice Hall, New York.

Hartman, D.E. & Lindgren, J.H. Jr. (1993). Consumer Evaluations of Goods and Services - Implications for Services Marketing. *Journal of Services Marketing*, 7 (2), 4 - 15.

Henly, S. J. (1993). Robustness of some estimators for the analysis of covariance structures. *British Journal Of Mathematical And Statistical Psychology*, 46(2), 313-338. doi:10.1111/j.2044-8317.1993.tb01019.x

Hirschman, E. C., & Holbrook, M. B. (1982). Hedonic Consumption: Emerging Concepts, Methods and Propositions. *Journal Of Marketing*, 46(3), 92-101.

Hoch, S. J. (2002). Product Experience Is Seductive. *Journal Of Consumer Research*, 29(3), 448-454.

Hoch, S. J., & Deighton, J. (1989). Managing What Consumers Learn from Experience. *Journal Of Marketing*, 53(2), 1-20.

Holbrook, M. B. (2007). Priceless: Turning Ordinary Products into Extraordinary Experiences. *Journal Of Macromarketing*, 27(1), 90-94. doi:10.1177/0276146706296957

Holbrook, M. B., & Hirschman, E. C. (1982). The Experiential Aspects of Consumption: Consumer Fantasies, Feelings, and Fun. *Journal Of Consumer Research*, 9(2), 132-140.

Hu, L. T & Bentler, P. M. (1995). Evaluating model fit. In: R. H. Hoyle (Ed.), Structural Equation modeling:Concepts, issues and applications (pp. 76-99). Sage, Thousand Oaks CA.

Hueiju, Y., & Wenchang, F. (2009). Relative impacts from product quality, service quality, and experience quality on customer perceived value and intention to shop for the coffee shop market. *Total Quality Management & Business Excellence*, 20(11), 1273-1285. doi:10.1080/14783360802351587

Jacoby, J. and Chestnut, R. (1978). Brand Loyalty: Measurement and Management. John Wiley & Sons, New York.

Jones, E., & Mustiful, B. W. (1996). Purchasing behavior of higher- and lower-income shoppers: A look at breakfast cereals. *Applied Economics*, 28(1), 131.

Jones, T. O. (1996). Why satisfied customers defect. *Journal Of Management In Engineering*, 12(6), 11.

Jöreskog, K.G. (1999). Some contributions to maximum likelihood factor analysis. *Psychometrika, 32*, 443–482.

Junhong, C., Chintagunta, P., & Cebollada, J. (2008). A Comparison of Within-Household Price Sensitivity Across Online and Offline Channels. *Marketing Science*, 27(2), 283-299.

Klaus, P. & Maklan, S. (2011). EXQ: a multiple-scale for assessing service experience. *Journal of Service Management*, 23, 1.

LaSalle, D. & Britton, T.A. (2003) Priceless: Turning ordinary products into extraordinary experiences, Harvard Business School Press, Boston.

Lax, H. (2012). The Experience –Loyalty-Value Connection. *Marketing News*, 10, 24-27

Lei, M., & Lomax, R. G. (2005). The Effect of Varying Degrees of Nonnormality in Structural Equation Modeling. *Structural Equation Modeling*, 12(1), 1-27. doi:10.1207/s15328007sem1201_1

Lemke, F., Clark, M. & Wilson, H. (2010). Customer experience quality: an exploration in business and consumer contexts using Repertory Grid Technique. *Journal of the Academy of Marketing Science*.

Macdonald, E., Martinez, V. & Wilson, H. (2009) Towards the assessment of the value-in-use of product-service systems: a review. Performance Association Conference. Dunedin, New Zealand.

Maklan, S., & Klaus, P. (2011). Customer experience. *International Journal Of Market Research*, 53(6), 771-792.

Malhotra, N. (2010). Marketing Research: An Applied Orientation (6th ed.). Pearson, London.

Maslow, A., & Lowery, R. (Ed.). (1998). Toward a psychology of being (3rd ed.). Wiley & Sons, New York.

Maslow, A.H. (1987). Motivation and personality (3rd ed.). Harper Collins, New York.

McDonald, R.P. (1985). Factor Aanalysis and Related Methods. Erlbaum, Hillsdale.

Merz, M. A., Yi, H., & Vargo, S. L. (2009). The evolving brand logic: a service-dominant logic perspective. *Journal Of The Academy Of Marketing Science*, 37(3), 328-344. doi:10.1007/s11747-009-0143-3

Meyer, C., & Schwager, A. (2007). UNDERSTANDING CUSTOMER EXPERIENCE. *Harvard Business Review*, 85(2), 116-126.

Michelli, J. A. (2007). The Starbucks Experience: 5 Principles for Turning Ordinary into Extraordinary. McGraw-Hill, New York.

Mulaik, S.A., James, L.R., Van Alstine, J., Bennett, N., Lind, S. & Stilwell, C.D. (1989). Evaluation of goodness-of-fit indices for structural equation models. *Psychological Bulletin, 105*, 430–445.

Nakada, P. (2005). Accounting for Risk. *Journal Of Performance Management*, 18(2), 44-55.

Oliver, R. (1999). Value as excellence in the consumption experience. In M. Holbrook (Ed.), Consumer value: A framework for analysis and research (pp. 43–62). Routledge, London.

Parasuraman, A. A., Zeithaml, V. A., & Berry, L. L. (1988). SERVQUAL: A Multiple-Item Scale for Measuring Consumer Perceptions of Service Quality. *Journal Of Retailing*, 64(1), 12-40.

Patrício, L., Fisk, R. P., Falcão e Cunha, J., & Constantine, L. (2011). Multilevel Service Design: From Customer Value Constellation to Service Experience Blueprinting. *Journal Of Service Research*, 14(2), 180-200. doi:10.1177/1094670511401901

Pauwels, K. & Reibstein, D. (2010). Challenges in Measuring Return on Marketing Investment: Combining Research and Practice Perspectives, in Naresh K. Malhotra (ed.), *Review of Marketing Research, Volume 6)*, pp.107-124.

Payne, A. F., Storbacka, K., & Frow, P. (2008). Managing the co-creation of value. *Journal Of The Academy Of Marketing Science*, 36(1), 83-96.

Peltonen, T. A., Sousa, R. M., & Vansteenkiste, I. S. (2012). Wealth Effects in Emerging Market Economies. *International Review Of Economics And Finance*, 24(1), 155-166. doi:http://dx.doi.org/10.1016/j.iref.2012.01.006

Pine II, B.J., & Gilmore, J. H. (1998). WELCOME TO THE EXPERIENCE ECONOMY. *Harvard Business Review*, 76(4), 97-105.

Pine II, B.J., & Gilmore, J. H. (2011). The Experience Economy: Work Is Theatre & Every Business a Stage (2nd ed.). Harvard Business School Press Books, Boston.

Poulsson, S. G., & Kale, S. H. (2004). The Experience Economy and Commercial Experiences. *Marketing Review*, 4(3), 267-277.

Prahalad, C. K., & Ramaswamy, V. (2004). CO-CREATION EXPERIENCES: THE NEXT PRACTICE IN VALUE CREATION. *Journal of Interactive Marketing*, 18(3), 5-14.

Reichheld, F. F. (1993). Loyalty-based management. *Harvard Business Review*, 71(2), 64-73.

Reichheld, Frederick F. (2003). The one number you need. *Harvard Business Review*, 81(12) 46-54.

Reynolds, K. & Arnold, M. (2000). Customer loyalty to the salesperson and the store: Examining relationship customers in an upscale retail context. *Personal Selling Sales Management*, 20(2), 89-99.

Rindskopf , D. (1984). Structural Equation Models: Empirical Identification, Heywood Cases, and Related Problems. *Sociological Methods & Research*, 41(1), 109-19.

Ross, J., Irani, I., Silberman, M. Six, Zaldivar, A., and Tomlinson, B. (2010). Who are the Crowdworkers?: Shifting Demographics in Amazon Mechanical Turk. In: CHI EA 2010. (2863-2872).

Rust, R. T., Moorman, C., & Dickson, P. R. (2002). Getting Return on Quality: Revenue Expansion, Cost Reduction, or Both?. *Journal Of Marketing*, 66(4), 7-24.

SAS Study Shows Customer Experience Is a Competitive Differentiator. (2009). *Teller Vision*, (1379), 5.

Satorra, A., & Bentler, P. M. (2001). A scaled difference chi-square test statistic for moment structure analysis. *Psychometrika*, 66, 507-514.

Schiffmann, L.G. & Kanuk, L.L. (2000). Consumer Behavior (7th ed.). Prentice Hall, New Jersey.

Schmitt, B. H. (1999). Experiential Marketing. Free Press, New York.

Schmitt, B. H. (2003). Customer Experience Management. John Wiley & Sons, New York.

Selnes, F., Gonhaug, K. (2000). Effects of supplier reliability and benevolence in business marketing. *Journal Of Business Research*, 49(3) 259-270.

Shankar, V., Berry, L. L., & Dotzel, T. (2009). A Practical Guide to Combining Products + Services. *Harvard Business Review*, 87(11), 94-99.

Shaw, C. (2002). The DNA of customer experience: How emotions drive value. Palgrave, New York.

Sobel, M. E., & Bohrnstedt, G. W. (1985). Use of null models in evaluating the fit of covariance structure models. In N. Tuma (Ed.), Sociological methodology 1985 (pp. 152-178). Jossey-Bass, San Francisco.

Song, J., Chermack, T. J., & Kim, H. (2008). Integrating Individual Learning Processes and Organizational Knowledge Formation: Foundational Determinants for Organizational Performance. *Human Resource Development Review*, 11 (4), 23-56.

Sweeney, J. C., & Soutar, G. N. (2001). Consumer perceived value: the development of a multiple item scale. *Journal of Retailing*, 77, 203–220.

Tellis, G. (1988). Advertising loyalty, exposure and brand purchase: A two-stage model of choice. *Marketing Research*, 25(2), 134-145.

Ting-Yueh, C., & Shun-Ching, H. (2010). Conceptualizing and measuring experience quality: the customer's perspective. *Service Industries Journal*, 30(14), 2401-2419. doi:10.1080/02642060802629919

Trigg, A. (2004). Deriving the Engel Curve: Pierre Bourdieu and the Social Critique of Maslow's Hierarchy of Needs. *Review Of Social Economy*, 62(3), 393-406. doi:10.1080/0034676042000253987

Tynan, C., McKechnie, S., & Chhuon, C. (2010). Co-creating value for luxury brands. *Journal Of Business Research*, 63(11), 1156-1163. doi:10.1016/j.jbusres.2009.10.012

Ulaga, W., & Reinartz, W. (2011). Hybrid Offerings: How Manufacturing Firms Combine Goods and Services Successfully. *Journal Of Marketing*, 75(6), 5-23. doi:10.1509/jmkg.75.6.5

Vargo, S. L., & Lusch, R. F. (2004). Evolving to a New Dominant Logic for Marketing. *Journal Of Marketing*, 68(1), 1-17.

Verhoef, P. C., Lemon, K. N., Parasuraman, A. A., Roggeveen, A., Tsiros, M., & Schlesinger, L. A. (2009). Customer Experience Creation: Determinants, Dynamics and Management Strategies. *Journal Of Retailing*, 85(1), 31-41. doi:10.1016/j.jretai.2008.11.001

Wahba, M. A., & Bridwell, L. G. (1976). Maslow Reconsidered: A Review of Research on the Need Hierarchy Theory. *Organizational Behavior & Human Performance*, 15(2), 212-240.

Yuan, K. H., Chan, W., & Bentler, P. M. (2000). Robust Transformation with Applications to Structural Equation Modeling. *British Journal of Mathematical and Statistical Psychology*, 51 (1), 31-50

Zeithaml, V. (1988). Consumer perceptions of price, quality, and value: a means-end model and synthesis of evidence. *Journal of Marketing*, 52, 2–22.

Zeithaml, V. A., Berry, L. L., & Parasuraman, A. (1996). The behavioral consequences of service quality. *Journal of Marketing*, 60, 31–46.

Appendix

In order to support legibility and practicability, the tables you can find throughout this study are a summary of the original outputs from SPSS and AMOS. To avoid repetition and to keep the page number to a minimum, the appendix merely contains the scales, the survey text, and the multi-group-effects for the structural models affected by the Heywood cases. The original outputs, models and further information regarding the implementation of this study are available upon request. You can contact the author via d.gurski@student.maastrichtuniversity.nl

Table A1. Initial CXQ scale items

Variable	Item	Measurement
	When I think of Starbucks, I recognize that …	{1,Strongly disagree} {2,Disagree} {3,Somewhat disagree} {4,Neither agree nor disagree} {5,Somewhat agree} {6,Agree} {7,Strongly agree}
PQ1	The quality of coffee is very good	
PQ2	The taste of coffee is very good	
SQ1	The service is very good	
SQ2	The employees are friendly	
SQ3	Starbucks does its best to satisfy me	
ATM1	The store design is appealing	
ATM2	I like the design of Starbucks stores	
ATM3	Starbucks stores have a nice atmosphere	
IMG1	The decoration of the store tells a story	
IMG2	Visiting Starbucks is a temporary escape from everyday life	
IMG3	I can escape into an imaginative world at Starbucks	
CON1	Time passes without notice when I am at Starbucks	
CON2	I often spend more time than initially intended at Starbucks	
SRP1	Starbucks surprises me	
SRP2	Starbucks delights me	
OP1	I usually meet friends or relatives at Starbucks	
OP2	I like to watch other people when I am at Starbucks	
OP3 *	I often feel disturbed by other customers	
COG1	I gain information about products while shopping	
COG2	I can easily accomplish my shopping goals there	
COG3	I learn something new while staying in this store	
AFF1	I enjoy being at Starbucks	
AFF2	I have fun at Starbucks	
AFF3	People can enjoy themselves at Starbucks	

* = negatively worded item

Table A2. Initial Perceived Wealth scale items

Variable	Item	Measurement
	When I think of my financial situation, I recognize that …	{1,Strongly disagree} {2,Disagree} {3,Somewhat disagree} {4,Neither agree nor disagree} {5,Somewhat agree} {6,Agree} {7,Strongly agree}
PW1	I perceive myself as wealthy	
PW2	I can afford a good life with the money I have	
PW3 *	I have financial problems	
PW4 *	I can only afford to buy things which are really necessary	

* = negatively worded item

Table A3. Initial Perceived Value scale items

Variable	Item	Measurement
PV1	I consider Starbucks' products a good buy	{1,Strongly disagree} {2,Disagree} {3,Somewhat disagree} {4,Neither agree nor disagree} {5,Somewhat agree} {6,Agree} {7,Strongly agree}
PV2	Starbucks' value for money is …	{1,very poor} {2,poor} {3,somewhat poor} {4,neither good nor poor} {5,somewhat good} {6,good} {7,very good}
PV3	Starbucks' prices are …	{1,very unacceptable} {2,unacceptable} {3,somewhat unacceptable} {4,neither acceptable nor unacceptable} {5,somewhat acceptable} {6,acceptable} {7,very acceptable}

Table A4. Initial Customer Loyalty scale items

Variable	Item	Measurement
WOM1	I say positive things about Starbucks to other people	{1,Extremely unlikely} {2,Unikely} {3,Somewhat unlikely} {4,Neither likely nor unlikely} {5,Somewhat likely} {6,Likely} {7,Extremely likely}
WOM2	I recommend Starbucks to someone who seeks my advice	
WOM3	I encourage friends and relatives to go to Starbucks	
WOM4	I consider Starbucks as my first choice to buy coffee	
WPM1	I continue to do business with Starbucks if their prices increase somewhat	
WPM2	For the benefits I currently receive at Starbucks I pay a higher price than competitors charge	

Figure A5. Published HIT with survey link in MTurk

Answer a 3 minutes survey about Starbucks

We are conducting an academic survey about Starbucks. We need to understand how you perceive shopping at their stores. Therefore you must have already shopped at Starbucks before.
Select the link below to complete the survey. At the end of the survey, you will be provided with a code to paste into the box below in order to receive credit for your participation.

Survey link: http://marketing.survalyzer-survey.maastrichtuniversity.nl/nq.cfm?q=24720c14-8356-4e6e-a984-eb41bb6dc65a

Provide the survey code here:

[Submit]

Table A6. Multi-group effects of Perceived Wealth in Structural Model #1

Multi-group effects	allPW	lowPW *	midPW	highPW	effect PW	support
$\beta_{PQ \to PV}$	.237	.447	.201	(-010)	-	H2*, H4*
$\beta_{CXQ \to PV}$	.704	.444	.818	.914	+	H1*, H3*
$\beta_{PV \to LI}$	.884	.808	.893	.916	+	H1*, H2*
$r_{PQ \leftrightarrow CXQ}$	.699	.596	.610	.837	+	

* Heywood case, i.e. inadmissible solution because e5 shows negative variance.
( ) insignificant effect for p=0.05

Table A7. Multi-group effects of Perceived Wealth in Structural Model #2

Multi-group effects	allPW	lowPW *	midPW	highPW	effect PW	support
$\beta_{PQ \to PV}$	.200	.470	(.105)	(-.070)	-	[H2*],H4*
$\beta_{CXQ \to PV}$	.696	.371	.842	.941	+	[H1*], H3*
$\beta_{PV \to LI}$	.383	.467	(.069)	.492	?	
$\beta_{PQ \to LI}$	.175	(-.015)	.362	.209	?	
$\beta_{CXQ \to LI}$	.415	.489	.591	(.289)	?	
$r_{PQ \leftrightarrow CXQ}$	.706	.610	.612	.838	+	

* Heywood case, i.e. inadmissible solution because e5 shows negative variance.
( ) insignificant effect ; [ ] Hypothesis is partially supported

Table A8. Multi-group effects of Perceived Wealth in Structural Model #3

Multi-group effects	allPW	lowPW *	midPW	highPW	effect PW	support
$\beta_{PQ \to LI}$	.133	(.067)	.363	(-.107)	?	
$\beta_{CXQ \to LI}$	.815	.835	.658	1.040	?	
$r_{PQ \leftrightarrow CXQ}$	.735	.656	.614	.875	?	
$r_{PQ \leftrightarrow PV}$	.691	.694	.621	.718	?	
$r_{CXQ \leftrightarrow PV}$	.894	.786	.911	.928	+	

* Heywood case, i.e. inadmissible solution because e5 shows negative variance.
( ) insignificant effect